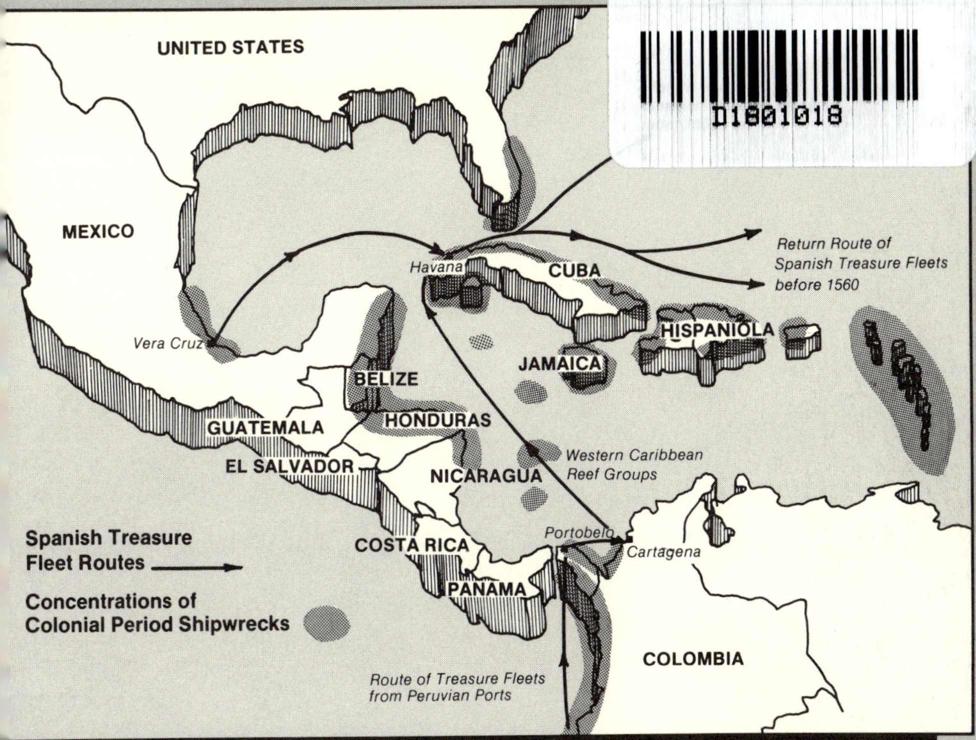

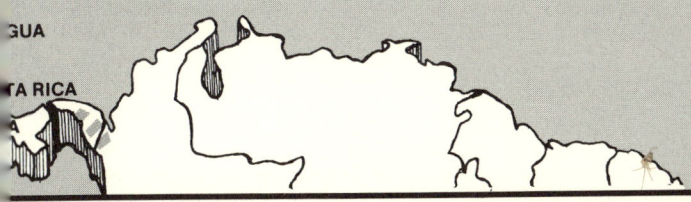

Gold-Bearing Areas of North America

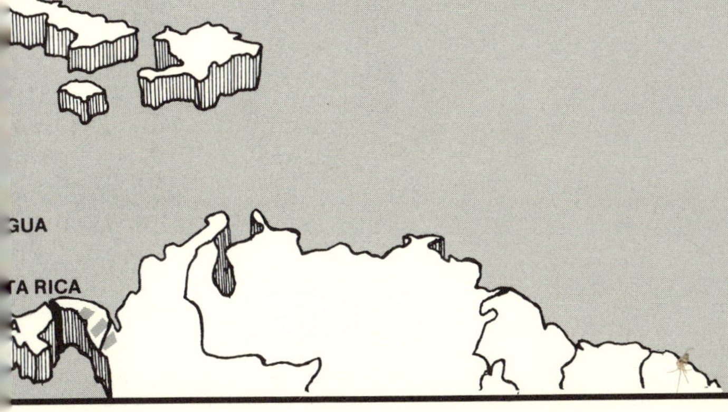

In Search of Gold

IN SEARCH OF GOLD

- Rock Mining
- Gold Panning
- Treasure Hunting
- Coin Beachcombing
- Artifact Excavating

STEPHEN M. VOYNICK

PALADIN PRESS
BOULDER, COLORADO

Hell, since a man's gotta dig for somethin'—
it may as well be gold.

—*anonymous Alaskan prospector*

Color photographs by Stephen M. Voynick.
Artifact photos courtesy of Museo del Oro.
Book and dustjacket design by
Janet Snow Ritchie and David Bjorkman.

In Search of Gold
Copyright © 1982 by Stephen M. Voynick

All rights reserved. Except for use in a review,
no portion of this book may be reproduced in any
form without the express written permission of the
publisher. Neither the author nor publisher assumes
responsibility for the use or misuse of any information
contained in this book.

First Edition
Published by Paladin Press, a division
of Paladin Enterprises, Inc., P.O. Box 1307,
Boulder, Colorado 80306
Printed in the United States of America

Library of Congress Cataloging in Publication Data
Voynick, Stephen M.
    In search of gold.
    1. Gold mines and mining.   2. Gold.   I. Title.
TN420.V68   622'.3422   81-16900
ISBN 0-87364-238-4   AACR2

THIS BOOK IS FOR BETTY AND JOE.

CONTENTS

INTRODUCTION: THE ETERNAL QUEST   1

I: FROM THE AGES   11

1: Old World Heritage   13
2: Gold by Conquest   19
3: Gold by Exploitation   29
4: Gold and the Common Man   37
5: And Still More Gold   47
6: Out of Sight, Out of Mind   57
7: Gold Fever Again   63

II: FROM THE ROCK   69

8: Mountains of Gold   71
9: Placer Gold   77
10: Lode Gold   83
11: New Technology   87
12: New Profits   95
13: In the Gold Fields   101
14: New Gold Rushes   109

| III: | FROM THE SEA | 117 |

15:	Sea-Going Gold	119
16:	Early Salvors	129
17:	Salvage Technology	133
18:	Modern Treasure Salvage	141
19:	On Reefs of Gold	151
20:	East Coast Coin Beaches	159

| IV: | FROM THE GRAVES | 163 |

21:	The Gold of El Dorado	165
22:	Golden Touches	171
23:	Professional Graverobbers	177
24:	Legalities and Realities	185
25:	The Graves	191

INTRODUCTION
THE ETERNAL QUEST

Twenty thousand years ago, in the latter part of the Stone Age, an early man plodded along a nameless stream seeing only the sticks and stones of his still-precarious existence on earth. Abruptly he halted, his eyes fixed upon a small but unusually bright object lying in the gravel at his feet. Bending, he plucked the curious piece from the stream bed. Then, turning it slowly in his fingers so as to catch and reflect the sunlight, he became fascinated by its rich, yellow luster. Hefting the irregular object in his hand, he found it to be far heavier than he expected for its size. Dimly realizing that he possessed something extraordinary, he again sank to his knees, this time to sort in earnest through the sticks and stones, seeking more of the heavy, yellow pieces. Man had begun his eternal quest for gold.

As still other unique properties of the yellow metal—it was so easily workable, yet virtually indestructible—became known, its appeal increased. In the milleniums that brought man into Neolithic times, gold evolved from a simple collectible to the most desirable natural material in the world. Based on both desirability and rarity, gold took on a recognized value which, in time, became the ultimate yardstick against which the value of all other materials, as well as the wealth and power of people and nations, were measured. Gold was simultaneously rare, beautiful, and eternal. Man eagerly made it the focus of his dreams, the goal of his adventures, and the universal symbol of wealth and power. So exalted was gold that the search for it became every bit as noble and enduring as the metal itself, driving gold hunters to deeds both glorious and ignominious. The quest for gold would inspire human beings far beyond the normal limits of courage and endurance, and would cause bloody wars to be fought, great civilizations to be destroyed, and entire nations to be built and broken.

Not even the passage of time would diminish the magical appeal, for even today we are just as irrational, obsessive, and fanatical in our relationship with the yellow metal we call *gold*.

Many of us now own gold in one form or another and, quite frankly, would like to acquire more. The acquisition of gold, to most persons today, automatically infers the purchase of coins, bullion, or jewelry from a conventional market source and at the prevailing market price. Ten years ago such acquisitions proved to be wise investments. But today, gold has already completed its skyrocketing adjustment in price and future increases will almost certainly be at far more conservative rates. Its purchase may not be considered a fast bargain at all, but rather another long-term commodity investment.

But there *is* another way to acquire gold today, and more individuals are engaged in it right now than at any time since the days of the Klondike in the late 1890s. The acquisition of gold from its natural sources represents one of the greatest bargains in history—a bargain based on a *profit potential* that has increased nearly as much as the price of gold itself. The profit a gold seeker realizes today is like most other business profits: the difference between costs and monies received from sale of the product, or in this case, sale of recoveries. Any other similarity, however, is not valid, for the "product" of the gold seeker is gold itself. Unlike other businesses in which profit is measured in paper dollars of steadily diminishing purchasing power, the gold seeker takes his reward directly in gold—the same material against which the dollar has declined so dramatically. Yet his costs, like other businesses', are paid in paper dollars. The profit potential enjoyed by a gold seeker today, thanks to the diminishing value of the dollar and the all-time high price of gold, has never been greater.

The accompanying chart shows the trend over two centuries of the purchasing power of gold (income), expressed as the price of gold per ounce, against the U.S. commodity wholesale prices (cost). During the 1800s and early 1900s the price of gold, (relative to the general upward trend of wholesale prices which reflects inflation), maintained a margin that made gold mining economically feasible in most cases. In other words, most miners could be assured that their average production would represent a profit. During this entire period the price of gold had been controlled. This meant that a miner with a steady production had a fixed income. His profit, however, decreased as inflation made his costs greater. For a miner to realize a large profit at that time was dependent solely upon his ability to recover a large quantity of gold. In the 1930s the wholesale price equaled the relative dollar value of gold, meaning the purchasing value of paper currency was highly questionable. The reaction was the run on the banks to convert paper to gold. The resulting devaluation of the dollar raising the price of gold to thirty-five dollars per ounce provided a substantial increase in the profit margin enjoyed by a gold seeker and was accompanied by an immediate increase in gold production. By the 1960s inflation had once again caught up with the fixed price of gold, eroding the profit margin. Accordingly, gold production in the Americas fell off to extremely low levels. Inflation had driven the cost of seeking gold to the point where, in most cases, it no longer made economic sense.

The chart also shows the spectacular climb in both indexes during the period 1970-80. This period was marked by the highest sustained inflation rate in history. During that decade most costs doubled, including the costs of gold

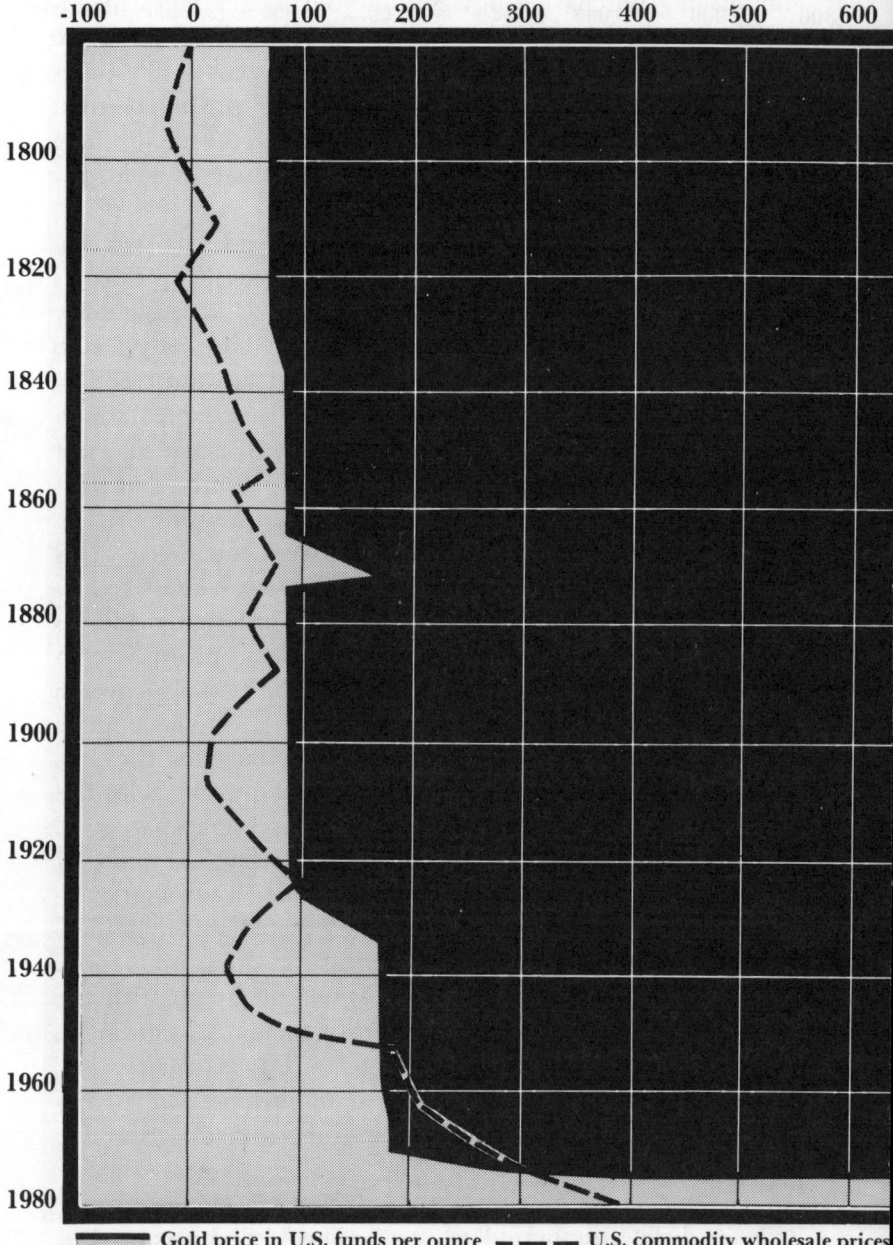

The margin between the price of gold index (the gold seeker's income) and the commodity price index (the gold seeker's cost)

R OF GOLD

800 900 1000 1100 1200 1300 1400 1500 %

may be considered the profit margin for an individual seeking gold from its natural sources. The extreme divergence of the two indexes in 1980 shows the tremendous increase in profit potential enjoyed by today's gold seeker.

hunting. The purchasing power of the paper dollar was cut in half. The price of gold, relative to the dollar, increased by a factor of *twenty*. In that economic arithmetic lies the reason why getting gold from its natural sources today is one of the greatest bargains in history.

Several other factors serve to increase even further that handsome profit potential. First, the quest for gold has always been a private and confidential matter. Exactly how much gold was recovered in a particular quest was known only by the individual who recovered it. Privacy and confidentialness are more important now than ever, in light of the very substantial tax liabilities on most other forms of incomes.

And second, many of the forms of gold recovered from natural sources command a premium value far above the current bullion value of fine gold. The worth of specimen nuggets, coins, artwork, and artifacts may be five or even *ten* times higher than the bullion value of gold.

Gold, the object of mankind's eternal quest, is certainly not a common metal, but neither is it the rarest, ranking fifty-eighth in abundance among the ninety-two natural elements found on earth. It has been assigned the chemical symbol *Au*, from *aurum*, the Latin word for gold which, in turn, was derived from *Aurora*, the goddess of the dawn, in respect to its sunlike color. Gold has an atomic weight of 197, a relative indication of element weight that is surpassed only by mercury (200), lead (207), and several other very rare, heavy metals. Platinum, considered the sister metal to gold, is a bit lighter with an atomic weight of 195. Iron, in comparison, has an atomic weight of only 55.8, less than one-third that of gold.

The density of gold, measured by specific gravity, is 19.3, meaning gold is exactly that many times heavier than an equal volume of water. Gold is five to eight times heavier than most common rocks which have specific gravities in the range of 2.5 to 4. Even silver, another highly valued metal, has a specific gravity of only 10.5, or a bit more than half that of gold.

An extremely soft metal, gold is rated at only 2.5 to 3

on the Mohs mineral hardness scale which places talc at 1 (softest) and crystalline carbon, or diamond, at 10 (hardest). Among the common metals, gold is considerably harder than lead, but somewhat softer than pure copper. Even with gold's great softness, the metal has a surprisingly high melting point of 1,063 degrees Centigrade, about the same as copper but far below that of iron.

While such technical measurements are interesting, they are hardly necessary for one to appreciate the great weight of gold. In the hand, the metal has an unmistakable feeling of authority that commands instant attention and respect.

Gold is by far the most malleable and ductile of all metals. Its remarkable capacity to be flattened and molded was one of the most exciting stimuli to man's early artistic efforts. Not even the most extreme hammering will irreversibly damage or destroy the metal. Perhaps it even thrives under the abuse. More hammering only produces the most delicate leaflike films to further tease the creative imagination. Gold may be hammered to a thickness of only 1/300,000 of an inch without deterioration, and a single ounce of the metal may be drawn into a continuous hairlike thread over sixty miles in length.

Another notable property of gold is its near inertness, or chemical inactivity, which accounts for its primary occurrence in nature as a native, or free, metal. Although capable of forming a limited number of chemical compounds, particularly as complex ores with the element tellurium, it is most commonly found in its bright yellow, metallic form. Gold never occurs pure, but always in combination with other metals, most often iron, copper, silver, or even platinum. Depending upon the percentage of the secondary metal present, the color will vary. The color of gold ranges from a silver-white, as with silver or platinum, to a warm reddish hue indicating the presence of copper. The same property of inertness is also responsible for gold's extraordinary durability. Early man soon learned that only the lesser metals were subject to corrosion and deterioration, while gold remained impervious to the effects of fire, water, or atmosphere.

It has only been in relatively recent times that man

learned that mercury, cyanides, and special acids could transform gold into other forms. Accordingly, virtually all the gold ever mined throughout history—about ninety-five thousand tons—is still in existence today. All that gold would make a solid cube only a mere thirty feet on a side.

Old World cultures elevated gold to its supreme position among metals, but it was in the New World, the Americas, that gold had its greatest effect on history and the common man. It was in the Americas that the *individual* quest for gold became a reality and reached its peak, its glamor and appeal founded upon the alluring fact that neither specialized knowledge nor substantial capital was required. Any man who could provide an investment in courage, muscle, and initiative, and who could accept the risks and consequences, could join the quest.

The great conquests and epic gold rushes are now history, and many believe the individual quest for gold has also faded into the past. Nothing could be further from the truth, for more gold than ever—with a higher value than ever—is being recovered by individual gold seekers right now. This renewed quest is not based upon a whimsical "golden dream," but on these solid facts of history, technology, and economy.

Fact 1: The recoverable gold remaining in the Americas today is greater than the total of all the gold ever taken by conquest, salvage, or mining.

Fact 2: The mechanical and electronic tools available now make today's gold seeker infinitely more effective and efficient than his predecessors.

Fact 3: Both the value and the profit potential, that difference between the real purchasing power of gold and the costs of the quest, have attained all-time highs.

As if these reasons were not enough, there are still other

features of the quest that many gold seekers today deem equally important. The quest for gold remains the most simple and direct approach to a livelihood as well as a possible fortune, pitting one against natural obstacles rather than the far more complex and frustrating problems of more conventional endeavors. The quest for gold is capitalism in its purest form and may be as free of regulation and restriction as one chooses to make it.

As always, adventure and the quest for gold are synonymous. The challenge of gold-getting will bring the individual as much true adventure as he or she is capable of embracing.

This is the story of the quest for the gold of the Americas, past and present. It is the story of how gold was, and is, taken from the rock, from the sea, and from the graves, and how history, technology, and economics have combined to make this present time—right now—the real golden age for the individual gold seeker.

There is little disagreement among market analysts that gold, over the long term, will surpass the one-thousand-dollar level. The only point they disagree upon is how soon this will happen.

I
FROM THE AGES

I
OLD WORLD HERITAGE

The first men to hold native gold in their hands were an extremely practical people. They *had* to be practical to cope with the overwhelming matter of simple survival. They had neither the time nor the interest to pursue matters not directly related to their survival. The gold they found, because of its softness, had no imaginable utilitarian value for use in tools or weapons.

JEWELRY AND GRAVEN IMAGES

The aesthetic qualities of the yellow metal, however, were more than sufficient to inspire the creation of man's first jewelry, the crude hand-hammered amulets recovered from a Stone Age cave in eastern Europe. Ornamental jewelry, as one would expect, became the first use of gold. But by Neolithic times man had bestowed upon gold a value significantly greater than its ornamental or decorative worth. This was an intrinsic quality, a universally recognized value based upon weight. The assignment of value reflected man's increasing veneration of gold. That veneration was somewhat irrational for, apart from its aesthetic appeal and intrinsic value, it was still an utterly impractical metal that would have no significant utilitarian use until modern times.

The special position of gold was already firmly established by the time the first high cultures emerged. Five thousand years before Christ, the Sumarians considered gold to be divine, using it extensively in their most sacred temples and also as royal insignia for their kings and rulers. The Sumarians did not produce gold, but acquired it through trade with other cultures in Afghanistan and India. The first test of the intrinsic value of gold took place on the lonely desert trails where bands of robbers constantly attacked the caravans to capture the shipments of the metal. When successful, the robbers would not hesitate to hammer down the most sacrosanct religious object into a form of gold more convenient to their own purposes. The appeal and value of gold would bridge cultures and religions. It was sought by both princes and peasants.

DEATH IN EGYPTIAN MINES

All of the gold used by the earliest high cultures was mined from placer deposits by the simple washing and sorting of loose stream gravels. The rising Egyptian civilization soon exploited lode, or hard-rock, deposits of the metal, constructing the first underground mines to help satisfy their prodigious requirements for art and religious use. In Akita, Egypt's eastern desert near the Red Sea, the Egyptians, relying only upon slave labor and without the advantages of any explosives or mechanical contrivances, engineered vast underground workings which even wound beneath the sea itself. Tunnels were advanced by the primitive method of heating the solid rock with open fires, then dashing it quickly with water. The resulting cracks enabled the rock to be broken and carried to the surface.

Countless thousands of slaves died in these mines, which were probably the worst the world has ever seen. Underground fires deprived the air of oxygen and released gaseous arsenic vapors from the rock. Slaves ordered to the mines survived only a few months, and those who somehow escaped

CHAPTER 1: OLD WORLD HERITAGE

suffocation or poisoning were killed by accidents and continual rock falls.

But gold—not the death of slaves—was the important consideration, and for centuries the Akita mines were a major source of Egyptian gold. An English company conducted underground mining operations from 1899 to 1921 in the same area, encountering "miles of old tunnels, an incredible number of human bones, and three million dollars in gold," (Gina Allen, *Gold*, 1964). This was the gold the Egyptians, four thousand years earlier, had been unable to recover themselves.

Egyptian contributions to gold production went beyond mining techniques. By 2,000 B.C. the Egyptians had devised practical procedures for smelting and refining the metal. Previously, gold could be recovered only if it occurred in particles large enough to be separated by washing from the gravel or crushed ore. The Egyptians constructed special furnaces capable of melting even the smallest gold particles from the rock and devised the first crude refinement process which employed simple evaporation and condensation to produce gold of fairly high purity.

FIRST COINS IN ANCIENT LANDS

About 1,000 B.C. man assigned a new role to his favored metal, this one of a formal monetary function. The Chinese began circulating gold in the form of small cubes of measured weight, and therefore of known value, which gained quick acceptance among merchants and traders. As gold came into increasing use as a medium of exchange among social and economic groups, and even among vastly different cultures, enduring standards of measurement evolved to facilitate the new monetary function.

The earliest standard of measurement was probably the *grain*, a unit of weight equal to what was considered to be a "standard" kernel of the local grain. One of the earliest practical measures was the Babylonian *shekel*, which was made up of about two hundred fifty grains, or approximately one-half ounce. By 700 B.C. the Greeks were using gold in a standard coin and, since that time, gold coins have been struck by virtually every government that ever existed as well as by thousands of private and religious organizations. It was inevitable that gold would become the supreme coinage and its acceptance as such reinforced its already ancient position as the supreme artistic medium.

Most cultures of the world, whether judged to be civilized or not by others, used gold to produce idols, charms, religious objects, and every other conceivable type of artwork. Through it all, the intrinsic value of gold would never be questioned or doubted. When the Greek tyrants, for example, decided to increase their circulating coinage, they did so at the expense of statues and ornaments that were melted down. Man had already learned that the form of gold was not that important—it was the gold itself that was valued.

The Roman Empire collected gold systematically for centuries, then redistributed it through squandering before its collapse. Much of that gold eventually reached the coffers of the growing Catholic Church, cast into statues for its altars, woven into vestments for its princes, and hammered into delicate gold leaf to illuminate its books and manuscripts.

HOW TO MEASURE TREASURE

Our modern system of gold weights was founded in the Middle Ages with the appearance of the Italian *florin* (48 grains) and the *Tower Pound* (5,400 grains). The Tower Pound remained the standard of England until replaced by the *Troy* system, which originated at an international trade fair held at Troyes, France, in the 1500s. The Troy system is now the international standard for all gold weights, with twelve Troy ounces to one pound, twenty pennyweight to one Troy ounce, and 24 grains to one pennyweight.

As the extreme softness of gold made it unsuitable for many uses, a number of alloys were created which reduced the gold content to provide much greater hardness and durability. And since gold was already honored by having its own weight system, it was only right that so noble a metal should also have its own system of purity measurement. The carat scale, a twenty-four-part system for measuring precious

CHAPTER 1: OLD WORLD HERITAGE   17

stones, came into use and "24 karat" became the universal term for fine, or pure, gold. The letter *k* distinguishes the measure of gold from that of precious stones. "14 karat" would indicate an alloy of fourteen parts pure gold and ten parts base metal. Like the origin of grain weight, the carat system was also based upon a food product, the carob seed, and our word form today is traced back to the Italian *carato* and the Arabic *qirat*.

MAGIC AND AMALGAMATION

During the Middle Ages the absence of discoveries of new sources of gold led to a general shortage which adversely affected every economy in Europe. The Catholic Church maintained its hold on most of the available supply and, in a reversal of the Greek approach, took the coin donations of the faithful, melted them down, and reworked the metal into sacred ornaments. With the European supply of gold drying up, alchemists were commissioned to discover or invent methods to create the metal. Some of the best contemporary minds were assigned to this noble task and given carte blanche by their gold-hungry kings and lords.

Much thought was given to the transformation of lead and silver into gold, but by no means were the efforts limited to metals. Whatever the imagination could conjure was tried, including such bizarre attempts as mixing eggs and saffron with a great deal of heat and prayer.

The hoped-for gold never materialized, of course, but neither was alchemy mere mumbo-jumbo. In the dark recesses of the alchemists' chambers was formed the groundwork of the true sciences, immediately resulting in improved manufacturing techniques for gunpowder and porcelain.

Spain operated the largest mercury mines in the world and the heavy, silvery liquid metal was a tantalizingly logical candidate for transformation to gold. Experiments revealed the phenomenon of *amalgamation,* the ability of mercury to absorb half its weight in gold, then to permit simple separation. Without the discovery of new sources of gold, the process of amalgamation remained a laboratory curiosity. But it would not remain so for long, for in a short time it would be used to help recover more gold than man could imagine.

2
GOLD BY CONQUEST

The European interest in gold was stirred a bit in the mid- and late 1400s through recurring accounts of the riches of distant Cathay and Japan and even of the Middle East. But with the exception of hopeful alchemy and desperate efforts to make exhausted mines productive again, most European countries waited passively and impatiently, contributing little toward developing a new source of gold. Only the carracks of Portugal circumnavigated Africa to reach India, returning not only with exotic spices and products, but also with respectable quantities of gold acquired through trading both in India and the west coast of Africa. This new source of gold brought hope to the darkened economies of Europe and virtually every exploratory voyage from the late 1400s on was looking for gold.

NEW WORLD OF GOLD

When the Italian navigator and cartographer Christopher Columbus was ready to test his theory of the existence of a shorter westward route to the gold of the Indies (then the term for the general collection of lands including India,

Cathay, and Japan), he naturally applied first to Portugal for assistance. When Portugal chose not to risk its precious resources on his radical ideas, the forty-year-old navigator turned to Spain in 1492.

Ferdinand and Isabella, rulers of a backward and bankrupt kingdom, in desperation proved receptive to Columbus's proposal. From the very beginning, the key word was *gold,* and the division of the gold Columbus would hopefully acquire was elaborately spelled out in the royal contract. In historical retrospect, it is somehow fitting that a substantial portion of the financial backing of the great voyage was provided by donation of Isabella's golden jewelry.

Columbus, after a six-week voyage which ranks as one of man's greatest achievements, did indeed discover the "Indies." The discovery brought some disappointment, however, for there were no gold-gilded palaces on San Salvador Island in the Bahamas, only uncivilized "Indians" running about naked. The single encouraging fact, as Columbus noted in his journal, was that they possessed simple ornaments and nose rings of gold. Columbus continued westward to Cuba, which he thought was Japan, then reversed his course to discover the island of Hispaniola. He lost his flagship, the *Santa Maria,* on a reef. He also discovered the first significant source of gold in the New World.

The Indians of Hispaniola freely traded their gold nuggets and ornaments for cheap European trinkets since gold, to them, had merely decorative value. Columbus misinterpreted their seeming generosity to mean that enormous quantities could be easily obtained in the interior. The loss of the *Santa Maria,* coupled with the certainty of gold, prompted Columbus to leave forty of his men at the settlement he named Navidad. There they could begin to accumulate the gold in preparation for his return the following year.

The two remaining ships continued east making a final stop at Samana Bay, in what is now the Dominican Republic, where he acquired a bit more gold. He also accomplished another first before sailing for Spain, engaging in an armed battle with the Indians.

Welcomed as a hero in Spain, Columbus had unknow-

CHAPTER 2: GOLD BY CONQUEST

ingly cast the mold for the coming Spanish adventure that would last over three hundred years: he had lost a ship, fought with the Indians, and acquired gold. The Spanish would repeat his performance a thousand times.

In 1493 Columbus encountered none of his earlier problems in mounting another voyage, thanks to the exaggerated reports of gold in the Indies. He now had seventeen ships under his command and thousands of men clamoring to join his crews, hoping to witness, and perhaps share in, the heaps of gold the Navidad crew had accumulated. Columbus discovered many more islands, but his return to Navidad brought only disappointment. The men of Navidad had acted overzealously in their demands for gold, becoming callous in their treatment of the Indians. They paid dearly for their indiscretions. The entire settlement, to a man, had been slaughtered by the natives.

Columbus soon learned that there were no rich mines on Hispaniola, and that the gold the Indians wore and traded was the accumulation of many years of slow production. The Spanish proceeded to establish a placer mining industry in the interior under the direction of some of Spain's most accomplished miners. Shipments of mercury for use in the amalgamation recovery process began arriving and each year a small fleet returned to Spain bearing the first eagerly awaited regular shipments of New World gold.

GOLD GOTTEN AT ALL HAZARDS

What most historians consider mankind's greatest adventure—the exploration and exploitation of the New World—had begun. Those historians also cite three factors as the forces motivating the Spanish to the conquest of a hemisphere: the gospel, personal glory, and gold.

Proliferation of the gospel played a major role in sustaining the great adventure, but too often religious ideals justified the atrocities the Spanish committed in their quest for the third factor. And while a degree of personal glory might be gained from a successful missionary effort or discovery of an unknown coast, the highest personal glory was reserved only for those who acquired substantial quantities of, once again, the third factor.

The third factor—gold—stood by itself. On the bottom line, it was the quest for gold that opened the Americas. King Ferdinand formally set the tone of future explorations when he issued his famed edict of 1511 to his men in the Indies: "Get gold, humanely if you can, but at all hazards, get gold."

The fortified city and harbor of Havana, Cuba, became a secure staging area for the Spanish gold seekers as they turned toward the Main, as they referred to the coasts of Central and South America, in search of a vague, ill-defined golden paradise, a New World version of the golden dream that man had always cherished. The men of Spain, teased by the modest amounts of gold that had already come from the Main, were absolutely certain of the imminent discovery of the golden paradise. Already the Indians of the South American coast had pointed inland, beyond the trackless jungles and unexplored mountains, and told fantastic tales of a city of gold, even of a "golden man." Both were manifest in the haunting name of *El Dorado*.

The intriguing tales of El Dorado brought only the promise of gold, not the metal itself, for a true golden fortune eluded the Spanish during the first quarter-century of their presence in the New World. Spain's first true bonanza came about through the fanatical obsession for gold on the part of Hernán Cortes. In 1519 Cortes, who already whetted his appetite on Cuban gold, was appointed to lead an expedition to conquer Mexico. His undisguised raw ambition, however, gave the Cuban governor second thoughts. The governor decided his own personal aims might be better served by choosing another man somewhat less headstrong. Cortes defied the last-minute order not to leave Havana and, with six hundred men supported by only sixteen horses and a few light cannon, sailed for Mexico. Landing near the site of Vera Cruz, Cortes dismantled his ships to discourage thoughts of desertion by his men, enlisted the aid of a thousand Cempoalan warriors who hated the Aztecs, and started marching inland. To the Indians, the Spaniards, with their horses and

CHAPTER 2: GOLD BY CONQUEST

frightening firearms, were white gods, a belief Cortes used to full advantage.

When Cortes arrived in Mexico, the Aztec Empire was thriving with a highly developed society and military backed by an extremely powerful religious order. The Aztecs had risen to power about the year 1200 and, two centuries later, had founded their capital city of Tenochtitlán very near the present site of Mexico City. Tribute, in the form of slaves and gold, was demanded and received from conquered tribes. Tenochtitlán became a veritable repository of gold, receiving two tons of the metal each year.

When Montezuma, the Aztec emperor, heard of the approach of the white gods, he sent messengers bearing gifts of gold to plead with them to venture no closer. Rather than having an appeasing effect upon Cortes, as Montezuma intended, the gold served only to fire the lust of the Spaniards. From that point the Aztec Empire was effectively doomed. Although ravaged by the effects of heat, cold, disease, and exhaustion, Cortes's columns trekked across some of the roughest terrain in all of Mexico. All the while, they received golden gifts from Montezuma.

The booty included a Spanish helmet filled to the brim with grains of gold, a two-hundred-pound golden sun disk as large as a cartwheel and inscribed with the symbols of the Aztec calendar, and an extravagant selection of gold statues and figurines. The gifts proved effective encouragement.

At Tenochtitlán, Cortes was welcomed by a confused Montezuma, who still thought the Spanish were gods and treated them with lavish hospitality. Cortes knew that, in reality, he and his men were captives in a golden prison, tremendously outnumbered. All escape from the city lay over Aztec drawbridges. With no other choice if all this gold was to be his, Cortes seized Montezuma and put him in chains as a hostage in his own palace. The huge ransom demanded in gold was brought to Cortes in the spring of 1520. Most of the ransom was in the form of exquisite artwork which the Spaniards unceremoniously cast into furnaces and melted down into bars.

Cortes, concerned now about the technical illegality of

his enormously successful venture, managed to reconstruct and dispatch a ship for Spain bearing the 20 percent crown tax to demonstrate his loyal intentions. When the governor of Cuba received word of this, he ordered a force of one thousand men to Mexico to relieve Cortes of his command. Leaving only eighty men in Tenochtitlán, Cortes marched to the coast where he surprised the force, captured the commander, and easily persuaded every man to join him with the honest promise of more gold than they had ever seen. But even as Cortes returned to the Aztec capital, the first of Spain's gold fell into other hands. The shipment of gold Cortes had dispatched to the King of Spain—the first true treasure shipment from the New World—was captured by French pirates.

When Cortes returned to the capital, he was greeted by the outbreak of open hostility as hordes of Aztec warriors attacked the Spaniards. Montezuma himself was stoned to death by his own people for continuing to urge moderation in dealing with the white gods. Collecting as much of their enormous accumulation of gold as possible, the Spaniards attempted escape at night over a secretly constructed drawbridge. In the ensuing battle, most of Montezuma's great golden treasure was lost into the lake.

The Spanish suffered very heavy losses, but those killed outright were far more fortunate than those captured who were doomed to a horrible slow death at the hands of the outraged Aztecs. Several hundred Spanish survivors, including Cortes himself, fended off constant Indian attacks to reach the coast.

Although only a small part of the original Aztec treasure ever reached Spain, it was more than enough to awe the Europeans. The few pieces of Aztec artwork that arrived intact drew lavish praise from the finest artists of Europe, who recognized technique and craftsmanship surpassing their own. Even so, the artwork was not spared from the furnaces and transformation into bars and coins. Virtually none of the golden artwork of the Aztecs survives today. Within two years of Cortes's setting foot into Mexico, Spain

CHAPTER 2: GOLD BY CONQUEST

found itself in possession of more gold than all the rest of Europe had accumulated.

CONQUEST OF THE INCAS

Even as the Spanish reveled in the gold that poured out of Mexico, another great Indian civilization still lived in peace along a three-thousand-mile section of the South American Pacific coast, unaware of the Spanish presence in their world. This was the Incan Empire, a highly developed society bonded together by worship of the sun, a common language, and an excellent road system that served for both transportation and communication. No culture in history had ever been so totally oriented to gold, nor did any have access to anything approaching the enormous quantities available to the Incas. By 800 B.C. the ancestors of the Incas had begun mining the western slopes of the Andes, which boasted the richest placers in the world. By the time the Incan civilization reached its zenith about A.D. 1400, gold mining was a highly organized and controlled industry. When the Spanish arrived, Incan production was something on the order of twenty tons annually, easily qualifying as the greatest gold production ever at that point in history.

Just as the Aztec Empire was toppled largely by the efforts of one man, so would the Incan Empire suffer a similar fate. This time the name would be Francisco Pizarro, another gold-obsessed Spaniard who, by 1531, was already an old hand at seeking the gold of the Americas. Eighteen years earlier, Pizarro had accompanied Balboa on his journey to the Pacific. Probably he had tortured the Indians to learn their gold source. What he had heard were tales of a golden civilization far to the south. The tales never left his mind. After two brief visits to the coast of Peru, he returned again in 1531 with a company of two hundred men exclusively intent on the conquest for gold.

That such a great empire could fall to one obsessed Spaniard and a band of poorly organized ruffians was due once again to the Indian belief that the intruders were some sort of gods. Pizarro wasted no time in capturing Atahualpa, the emperor, at the city of Cajamarca, killing two thousand Incas in the process.

With Atahualpa as hostage, the most incredible golden ransom in history was demanded, a ransom that would fill a chamber twenty-two by twenty-seven feet to a point as high as a man could reach. Journals record that, over several months, the huge ransom was borne to Cajamarca on over a hundred thousand litters. The actual weight of the ransom totaled some six tons. Pizarro himself considered the single most precious item to be Atahualpa's throne. The piece contained one hundred ninety pounds of gold, studded with clusters of emeralds and amethysts. Carrying the throne required twenty-five Incas. Just as happened in Mexico eleven years earlier, the golden artwork was melted down into bars. The Aztec treasure, which such a short time ago had seemed so fabulously rich to Spain, became a mere pittance in comparison to Pizarro's loot.

Atahualpa was strangled to death by the so-called gods. After his death the flow of gold stopped immediately; the incoming ransom was diverted to hiding places in lakes and mountain caves.

Small lots of the golden artwork have turned up since from time to time, and the remaining part is one of the richest prizes for treasure hunters today.

The loss of the remaining ransom caused Pizarro no great distress, for he was now able to focus his attentions on the city of Cuzco, the Incan capital, where even more gold was reportedly stored. A year after Pizarro took Cajamarca, his men entered Cuzco, without question the richest city in the New World. Although much gold had been removed before the anticipated arrival of the Spanish, quantities far in excess of Atahualpa's ransom remained. The Spaniards systematically looted the city, robbed the tombs and graves, defiled sacred mummies to remove their gold ornamentation, stripped over seven hundred golden plates from the walls of the Great Temple of the Sun, and carried away the magnificent golden masks and statues that adorned the interior.

CHAPTER 2: GOLD BY CONQUEST

When the Incan gold reached Spain, it represented by far the single greatest infusion of gold into any economy in history. About one hundred fifty *tons* of gold poured into a bankrupt nation that, only a few decades earlier, measured gold in mere ounces. Pizarro shipped back a collection of the most striking pieces of Incan artwork, including several near life-sized statues. After the usual brief period of admiration, every single piece was melted down into the bars and coins the Spanish considered the only practical form of gold.

In a few short decades Spain had gone from being bankrupt to being the richest nation in Europe. The national economy was tremendously stimulated, however artificially. The stimulated economy had a similar awakening effect upon literature, the arts, and the quality of Spanish life in general. It had all begun with the modest quantity of golden trinkets Columbus had found and had been sustained by the encouraging production of the Hispaniola placers.

Cortes's Aztec gold confirmed the golden promise of the New World, a promise that was fulfilled by the incredible golden fortune Pizarro took from the Incas. By 1540 not only was Spain the richest nation, it was also the largest nation on earth considering the vast extent of its New World empire. Although it was not comprehended at the time, Spain had also reached its economic, military, and political peak of power. Even though the gold would flow in for centuries, the Spanish Empire was already in decline and beginning the long road leading to its inevitable collapse, for the yellow metal it so deeply craved would prove both boon and bane.

3
GOLD BY EXPLOITATION

The plundering and destruction of the Aztec and Incan empires were the high points in terms of the acquisition of gold during the first fifty years of Spanish presence in the Americas. Only when the immediately available quantities of accumulated Indian gold were gone, did the Spanish begin working the rich placer deposits. The enormous richness of the mines together with the convenient availability of unlimited slave labor probably deterred the establishment of more organized and efficient techniques.

SYSTEMATIC SPANISH MINING

For a century most gold separation was accomplished by simple panning. The type of pan favored by the Spanish was the wooden *batea*, similar in shape to the gold pans of today. The bateas were fashioned from transverse cuts of timber chiseled out to hollowness and smoothed by rubbing with abrasive rock. The bateas functioned exactly as our modern pans by acting as a receptacle to hold and wash gravels while naturally retaining the heaviest materials. The largest bateas measured nearly four feet across and could wash a forty- or fifty-pound load of gravel. Panning with the bateas was a backbreaking job assigned to Indian slaves, who washed each load down to a small amount of concentrate, then turned it over to a Spanish overseer for final washing and collecting of the gold.

Only when the richest placer gravels had been worked did the Spanish turn to the lode deposits, the location of which required no great feat of prospecting. Most lode deposits were a short distance upstream from the placers and sometimes marked by a highly visible outcropping of white quartz.

In the beginning the Spanish used techniques not much advanced from those employed by the Egyptians three thousand years earlier. While the practice of heating the rock, then dousing it with cold water was still used, the Spanish also enjoyed the advantages of a wide variety of iron tools as well as gunpowder. A primitive form of drilling evolved in which an iron chisel was manually hammered into the rock, a slow, grueling task. The resulting shallow hole could then be loaded with gunpowder and ignited to fracture the immediately adjacent rock.

The narrow underground workings crept through the rock at a snail's pace. With the exception of the richest veins, the removed ore was simply broken quartz with the gold still locked in it.

Crushing the ore was first accomplished with hammers and later with the use of *arrastras,* crude crushing mills of stone comparable to a mortar and pestle arrangement. The mortar was constructed by hollowing out a natural stone base on a stream bank. A heavy, movable stone shaped to fit the mortar was the pestle. Power to drag the pestle around in a circle was provided first by slaves, then mules, and finally, in the case of more elaborate systems, with water. The product of the arrastras was a pulverized ore in which most of the gold was now free and could be recovered by washing in a batea.

Gold still locked in the quartz or too fine to separate by washing was then recovered through amalgamation. The demand for mercury was enormous. Almost the entire produc-

CHAPTER 3: GOLD BY EXPLOITATION

tion of the Almadén mercury mines in Spain, the greatest in the world, was shipped to the Americas for use in recovering gold. For centuries, a major facet of Spanish trans-Atlantic mining trade was the story of mercury being shipped west and gold being shipped east.

WILD CHASES FOR GOLDEN CITIES

In the years immediately after the conquests of the Aztecs and Incas, the primary Spanish concern was not about the systematic mining of gold, but rather where the next golden civilization would be found. The dream of finding it spurred some of the greatest explorations in history.

Francisco Coronado left Mexico in 1541 seeking the Seven Cities of Cibola, a North American El Dorado. Accompanied by three hundred Spaniards and one thousand Indians, Coronado marched north, first through Arizona, then into New Mexico where he did indeed find Cibola. This Cibola was not the expected city of gold, but the adobe cities of the Pueblo Indians. Coronado was then told of another golden city, this one called Quivera, a fictitious place obligingly created by the Indians who felt—indeed, they knew—it was what the Spaniards wanted to hear. Undertaking a monumental trek that brought him to the present state of Kansas and possibly even into Nebraska, Coronado found only disappointment on the vast plains. His return to Mexico was shrouded in failure as he had not found gold. Centuries would pass before his adventures were given rightful recognition as incredible feats of exploration.

Almost at the same time as Coronado pursued wild geese, Hernando de Soto led his large expedition out of Florida authorized by a license of conquest issued by the Spanish crown. In another difficult journey, he became the first European to see the Mississippi River and much of the southeast United States. De Soto found no gold either, but was spared the disgrace of returning in failure. He succumbed to disease, and his weighted body was secretly allowed to sink into the great river he had discovered lest the Indians realize the expedition's leader had died.

Meanwhile, in South America, the enticing legend of El Dorado was luring many into the rugged mountains and jungles, each hoping to emulate the grand accomplishments of Cortes and Pizarro. Leaving from points as far apart as Trinidad, on the Atlantic, and Peru, on the Pacific coast, six separate expeditions within ten years (1536-1546) took their chances in the steamy jungles and frigid mountain passes. All were generally aiming for a vague site somewhere near the present city of Bogotá, Colombia. Disease and exhaustion took their toll upon both the Spanish and the two German groups that operated under crown license.

The gold-working Muisca and Chibcha Indian cultures were discovered along with Lake Guatavita, the deep, mountain crater lake reputed to be the site where a mysterious "golden man" tossed great quantities of sacrificial gold into the depths. But the real El Dorado, as they conjured the place to be, was not found.

The Spanish did discover the enormously rich placer deposits of interior Colombia that would produce a steady supply of gold for the next two hundred years. By 1600 the Spanish were systematically working the placers, the last truly rich deposits they would have the pleasure of mining. Gold production was such that a mint was established at Bogotá, where the first gold coins in the New World were struck in the 1630s. The Colombian gold was notable not only historically and economically, but metallurgically as well.

CANNON OF PLATINUM

It was difficult to refine Colombian gold until 1735, when the Spanish succeeded in separating an unknown metal, one that was as heavy as gold, nearly as malleable, and which shone with a silvery luster. A few years later the metal was identified as a separate element and named platinum, from the Spanish *platina,* or "little silver." The Spanish deemed it totally worthless and even detrimental as it adulterated their beloved gold. In Colombia a great deal of the separated platinum was thrown into the streams below the Spanish mills and refineries.

CHAPTER 3: GOLD BY EXPLOITATION

In time, the Spanish used it as a hardening agent in the brass that was cast into cannon. Cannon have been recovered from shipwrecks in modern times containing as much as 5 percent platinum by weight. Since some of those guns weigh over fifteen hundred pounds each, and with the price of platinum today nearly double that of gold, the platinum content of each gun exceeds one million dollars' worth.

Platinum has always held a unique and interesting position in man's judgment relative to that of gold. The metal is every bit as "noble" as gold and even a bit more rare. The only detracting factor is its color. It is too close to silver. Even if platinum had somehow been discovered first, its "nobility" would only have been superseded by that of gold in time anyway.

FREE ENTERPRISE IN BRAZILIAN GOLD FIELDS

The next major gold bonanza fell not to Spain, but to Portugal. Settlement and exploration in the vast Portuguese colony of Brazil had, for one hundred fifty years, dragged on without the excitement of conquests and gold discoveries. While the presence of gold was known in the enormous trackless interior, an economy based upon coastal agriculture and the African slave trade continued to stagnate. By 1670 modestly encouraging placer deposits had been encountered. Then, in 1682, the fabulously rich placers of the Minas Gerais region were discovered. When word of the strike reached the coastal settlements, thousands of settlers abandoned their farms and ranches and, many with only packs on their backs, set off for the interior.

The Minas Gerais rush was significant not only because it produced a huge amount of gold, but also because it was the first rush in history where free enterprise played a major role. Portugal had a most difficult time attempting to establish order and collect its crown tax. For one reason, the gold fields were inaccessible. For another, a rampant individualism was fired by the possibility of making a personal fortune. New discoveries were made almost every day and the thousands of gold seekers who poured in represented the entire social strata of Brazil from escaped slaves and illiterate vaqueros to already wealthy noblemen.

For the first time in the Americas, the accepted currency used in the hastily constructed frontier towns was raw gold dust and nuggets. Violence, disease, and absurdly high prices for goods marked the chaotic rush. Many individuals did make personal fortunes. The patterns they established in the Brazilian gold fields would prove remarkably similar to the gold rushes that would take place in the western United States two hundred years later.

Three other major discoveries occurred in Brazil after the Minas Gerais rush. The rich placers at Cuibá were discovered in 1718; seven years later, more gold was found at Goiás; and in 1734 the great deposits at Guaporé were discovered deep in the Mato Grosso. Just as at Minas Gerais, thousands of hopeful gold seekers cut trails through jungles that had never been explored to reach the new gold fields. The Brazilian placers proved to be true bonanzas. By 1701 the Minas Gerais fields alone were producing two tons of gold each year. In 1740 the combined official production—that which was actually taxed—had reached sixteen tons per year.

Since government control was haphazard at best, a very capable smuggling industry was organized for the sole purpose of avoiding the crown tax. Sovereign loyalties and allegiances paled alongside the glitter of the gold, and it was a rare individual who chose to sell to Portugal and pay the tax when German, English, and Dutch interests stood ready to pay premium prices for the gold. For every ton of gold imported legally into Lisbon, another ton disappeared into the gold-running channels. Accordingly, the actual production of the Brazilian gold fields during the eighteenth century is unknown, but it probably easily exceeded one thousand tons.

Spain and Portugal exploited the gold fields of the Americas for three centuries, shipping as much as two thousand tons of gold to Europe. This was more gold than the world had ever known, and the ultimate effect upon the economies of both nations was utterly disastrous. Neither nation

CHAPTER 3: GOLD BY EXPLOITATION

ever built the domestic industries that would generate income internally but used the gold simply as payment for foreign imports. All of Europe enriched itself on the gold of the Americas, while Spain and Portugal relied only upon the next arriving shipment to sustain their hollow economies.

By 1800 the overstretched and undermaintained New World empires of both Spain and Portugal were swept by a wave of militaristic nationalism which, in only two decades, successfully overthrew the colonial yokes. While an incredible amount of gold was shipped from the New World, Spain and Portugal lost and left many, many times more. Aside from the great natural deposits and Indian graves which were barely touched, they left a trail of gold-laden shipwrecks as long as the shipping routes themselves. On land they left a trail of rusted swords and helmets to mark thousands of crude hard-rock mines and half-worked placers that stretched six thousand miles from the southwestern United States far into South America. And for the inhabitants of the newly independent American nations, they left a legacy of lust for gold that would inspire not the sovereign governments, but rather the individual gold seekers. As the last of the Spanish and Portuguese treasure ships departed for Europe, the age of gold and the common man was dawning in the Americas.

4
GOLD AND THE COMMON MAN

The final collapse of Spain's New World empire in 1820 found over a dozen sovereign nations occupying the hemisphere that had once been controlled and exploited by a handful of European powers. South and Central America was now a long line of independent republics, and North America was occupied by the United States from the east coast to the Rockies, Mexico in the Southwest and California, Russia in Alaska, and Britain in Canada. While sovereign claims represented quasi-legal ownership, the far western area of what is now the United States, that unpopulated and unexplored region from the Rockies to the Pacific, was really free for de facto occupation.

GOLD IN THE APPALACHIANS

Although gold is not usually associated with the history of the eastern United States, the first English settlers made a substantial effort to discover it. Captain John Smith, in 1608, sent samples of what he thought was gold ore to London for testing. It was not, and the hopes of discovering gold in the East began to gradually fade even though trace gold was found later in Appalachian streams. The Indians worked placers in several locations from Maine to Georgia but gold in exciting quantities was not found until 1799.

The first major United States discovery came in the form of a nugget reported to weigh an impressive seventeen pounds that was found in a North Carolina stream by a twelve-year-old boy. For three years the nugget, never positively identified as gold, was put to use as a doorstop, certainly one of the more expensive doorstops in history. Its eventual identification touched off the first United States gold rush. Over one hundred fifty pounds of gold were taken from streams near the discovery site and, for several decades, placer mining was conducted extensively. Production was great enough to warrant the construction of a United States mint at Charlotte.

The second Appalachian gold rush took place in 1829 in northeast Georgia where the discovery nugget was claimed to have been kicked up by a deer fleeing a hunter. By the end of the first year, over four thousand gold seekers had swarmed into the area and developed both the techniques and attitudes that would mark the coming western rushes. The basic procedures of prospecting with gold pans and washing with long sluice boxes were perfected. More importantly, it was clearly demonstrated in Georgia that the acquisition of gold in the United States would assume precedence over many "lesser" political and social considerations.

The gold discovery took place in the Cherokee Nation on territory which had been ceded to the Indians in perpetuity by the government of the United States "for as long as the grass grows and the river shall run." When the extent of the Georgia deposits became known, the grass stopped growing and the river stopped running abruptly. The government rescinded the treaty and banished the Cherokees to inferior lands including the dusty plains of Oklahoma.

The rush had peaked by 1837, but production again was great enough to construct another mint at Dahlonega, Georgia, to purchase the gold and mint it into coins. Over ten tons of gold were produced in the southeastern United States. Placer gold was valued then at about sixteen dollars

CHAPTER 4: GOLD AND THE COMMON MAN

per ounce, and the $5 million was a significant contribution to the neophyte U.S. economy.

MANIFEST DESTINY OR BUST

By 1840 Appalachian gold production had declined dramatically, but the United States had gotten its feet wet in the streams of gold, and many adventurous citizens liked it. The Oregon Trail, the first of the great westward migration routes, opened in the early 1840s, but the political and economic uncertainties of the era kept migration to a minimum. The land beyond the Rockies was still "Spanish America," and tales of hostile Indians and great, barren deserts offered more discouragement than promise.

Yet other forces were at work to make westward expansion inevitable. Social and economic conditions in the populated East were depressed, unemployment was high, and little hope of improvement in the foreseeable future was provided by a national economy that was lethargic and directionless. By 1845 the westward leaning was embodied in the philosophy of Manifest Destiny—the belief that God ordained the United States to stretch from Atlantic to Pacific. California, it was said, had rich soil, a favorable climate, and abundant water. There was even talk of gold in California, for placer deposits were first reported in 1820 and the first Mexican claim filed in 1842. Excitement about gold was slow to catch on, for California was still Mexican territory. Military successes against Mexico in 1846 served to reinforce the concept of Manifest Destiny but had also left the government financially drained. The nation desperately needed a major development in the West to encourage migration and settlement. If ever any people were ready politically, socially, and economically for a monumental gold strike, they were the people of the United States in the year 1848.

STRIKE AT SUTTER'S MILL

The golden event occurred on the twenty-fourth day of January of that year on the South Fork of the American

River. The discoverer was thirty-five-year-old James Marshall who found several nuggets in the millrace of a lumber mill he was building for a prominent local trader and farmer, John Sutter. Sutter was instrumental in playing down the discovery, fearing rightfully that a full-scale rush would trample his small, but promising empire. Even the first public pronouncement of the strike, a front-page article in the San Francisco *Californian* bannered "Gold Mine Found," was met with little enthusiasm.

Four months had passed before further newspaper mention, together with some well-orchestrated displays of the new-found gold by those who had already established stores in the discovery area, touched off the rush in San Francisco. Shops closed, sailors jumped ship, and farmers left their fields throughout central California to tramp for the gold fields. In May only a few hundred hopefuls panned the streams, but by the end of the year, their ranks had mushroomed to ten thousand.

The last half of 1848 was a confused and disorganized time along the American River. Still unknown was the true extent of the discoveries, and, thanks to the impatience of the early arrivals to see "gold in the pan," early mining was limited to just that—panning. The discovery deposits were so rich, however, that production totaled fourteen hundred pounds worth about three hundred thousand dollars.

Word of the strike reached every town in the Western Hemisphere. But nowhere was it received with more excitement than back East in the States, thanks largely to the press. Even though not even the Californians themselves knew the extent of the strike, the eastern press made fantastic claims during the 1848-49 winter. Stories stated that nothing less than "one thousand million dollars" was expected, and every day miners scraped out "hundreds of dollars" with nothing more than "their bare hands." Along with the newspapers appeared a host of hastily printed guidebooks, every one of them long on enthusiasm and horrendously short on fact. Whether or not the discovery warranted a rush, the United States needed one and the press assured it would get one. Manifest Destiny was now gold plated, and the trails westward were jammed in the spring of 1849. Most of the would-be gold seekers were either absurdly over- or underequipped, and only a few had a realistic geographic awareness of the trek they had undertaken. But all that mattered was the gold waiting in California. Those who could afford the considerable expense booked passage on ships sailing around Cape Horn. Others reached Panama, crossed the isthmus, then sailed north to San Francisco on coasters.

THE MOTHER LODE

During the first full year (1849) of the great California rush over one hundred thousand newcomers reached the gold fields. As the seekers fanned out over the western slope of the Sierras, the surprising extent and richness of the placer deposits were determined. They found gold all along a one-hundred-forty-mile section of the western slope, all of it originating in the great mother lode higher in the mountains. Some of the more fortunate individuals actually did stumble onto streams so rich that nuggets lay on the surface of the gravels. Many ten- and twenty-pound nuggets were found and most were celebrated in the press to sustain the excitement at fever pitch. Miles of pine forests were leveled almost overnight to provide lumber for cabins and the many miles of sluices and flumes that now lined every creek. By the end of 1849 even the most exaggerated press claims had come true. In one year the California gold fields had produced ten tons of gold worth about $5 million—equal to the entire total production of the Appalachian fields over fifty years.

VIGILANTISM AND INDIVIDUALISM IN THE FIELD

With little official government control to regulate the chaos, the miners managed admirably. Mining districts, a grand example of democracy bringing order from confusion, became the basis of organization in the gold fields. The miners drew upon the ancient Spanish-Mexican mining codes, claim limits, and recording processes to regulate their own operations. Claim jumping, both legal and physical, as well as violence and murder were unavoidable. Vigilante committees were formed to murder the murderers, restoring order and permitting the all-important mining of gold to continue.

Under the protection of their own laws, the gold seekers found themselves in a unique and very desirable situation. They were participating as equals in a major gold-seeking adventure in which the common man, the individual, owned every single grain of gold he washed from the gravel. There was no distant king claiming a crown percentage and no god demanding a sacrifice or offering. There was no landowner, either, to cut into the deal. Although the lands were the property of the United States, they had been declared "public domain" and open to any citizen for whatever use he chose. To acquire his own gold, a person needed no license and no permission, nor did he have to be a man of noble lineage with a title before his name. And when his work was done, he paid no taxes to any government. A man had only to reach the gold fields by whatever way possible, and to arrive with an ax, a shovel, and one of the metal pans that were now being manufactured by the tens of thousands. The ancient relationship of gold and man had reached a new era.

But many gold seekers—commoners and noblemen alike—were to reach only grim disappointment. Anyone reaching California after the spring of 1850 found every mile of every stream staked. If he would dig at all, he would do it as a hired laborer on another man's claim. The entire system of mining was also rapidly undergoing transition. Except on the smaller streams where it was impractical, the trend was toward company and large industrial operations backed by readily available eastern capital. As many as one hundred men could now be seen working one single rich claim, washing thousands of tons of gravel through a network of sluices and flumes every day. *Hydraulicking*—the use of powerful pumps to erode away twenty or even thirty feet of barren overburden to expose the gold-bearing layers—was employed extensively. By the end of 1850 nearly two hundred thousand persons, many bringing expertise from Mexico and Peru

CHAPTER 4: GOLD AND THE COMMON MAN

where gold mining was a centuries-old tradition, milled about the gold fields of California.

The land itself suffered terribly as ever more gold was torn out. Stream channels were permanently diverted and altered. Great heaps of boulders, laboriously pitched from the gravels, lined the banks of every stream. Forests were reduced to fields of decaying stumps. Once-clear water flowed thick with the mud and silt from ten thousand sluice boxes. And above it all, the once-quiet air was filled with the roar of the huge hydraulic pumps, the sound of the shovels, and the excited voices of the masses working harder than they had ever worked in their lives.

The words *California* and *gold* became synonymous quickly and for good reason. The 1851 production reached the staggering total of 2,500,000 ounces, just over one hundred tons, and was worth more than $50 million. That awesome production climbed still higher when one hundred sixty tons of gold, worth about $80 million, came out of the sluices. By 1855 the richest and easiest gravels had been worked; production declined sharply, but not before seven hundred fifty tons of gold, valued at $350 million, had been recovered. Production continued in California at a reduced rate, and, by 1875, a quarter-century after the great rush, production had risen to a total of twenty-five hundred tons of gold worth $1 billion. El Dorado had indeed been found, for that production was roughly equivalent to all the gold that had ever been mined in the history of the world.

POLITICS AND POETRY

The impact of California gold was felt everywhere. In the 1840s it was Russia that led the world in production, accounting for some 60 percent of the total. By 1851 Russia's gold-getting had fallen a distant second behind the United States'. The promise of Manifest Destiny had been largely fulfilled in two short years, and now a quarter-million Americans swarmed through newly acquired territories once considered vulnerable to foreign aims.

At the time of the actual discovery at Sutter's mill, California did *not* belong to the United States, but to Mexico. Had the Mexicans made any attempt to maintain control over California and the gold fields, there would have been a bloody chapter about it in the history books today.

As it happened, the U.S. achieved international financial credibility instantly. Much of the California gold was funneled to Europe to cover debts, and particularly to England and France which, for decades, had been reluctant to accept lesser forms of payment from the financially unsound United States. Gold coinage, both in the United States and Europe as well, increased ten times in a matter of a few years. Nearly every gleaming new coin had come from the California sluices.

The California rush had an enormous emotional and cultural impact upon the American people, instilling in them a new-found faith in both their personal and the national future. The Forty-Niners had proven the West was real and, gold or not, offered a viable alternative to life in the dreary East. California meant more than gold. California meant the future, individual betterment, and the achievement of a goal.

The Forty-Niners who made the great trek became an indelible part of American folklore with their adventures celebrated both in song—"I'm off to Californy with a washbowl on my knee"—and in the spoken word, as many of their colorful expressions found a niche in the Americanized English language. Sayings like "stake a claim," "strike it rich," and "see how it pans out," all had their origin in America's first great experience with gold.

America had liked that golden experience, loved it, in fact, even though in reality the golden fortunes went to the handful who "got there first," and even though that adventure was, and all future adventures would be, marked by hardship and risk. As much as gold itself, the quest for it became accepted as a worthy and noble American endeavor, but one clearly not for the weak-hearted or easily discouraged.

The American poet Edgar Allan Poe never saw the great California gold fields for, in 1849, he lay near death in the

CHAPTER 4: GOLD AND THE COMMON MAN 45

East. Yet, in one of his last works written a year earlier, he, too, had dreamed of El Dorado, had written of a "gallant knight" who must ride beyond the universe to find the treasure.

El Dorado was more than just a poet's imagining. El Dorado existed, of that there could be no question. The Spanish merely didn't recognize it when they found it. It had been found in Mexico, Peru, Colombia, Brazil, and then in California. And if any American felt he missed the chance to seek for it, he need only wait a few more years. In North America, just as in South America, El Dorados would come in bunches.

5
AND STILL MORE GOLD

Not all of the effect of the California gold upon the nation was good. The massive influx of gold into the economy spurred inflation which eventually led to a financial panic in 1857. Depression and unemployment worsened in the East. There were also the worries about the inevitable civil war drawing nigh. A drought in the Midwest drove up the price of food. And now that the gold fever had subsided in California, many of the original Forty-Niners began to drift back East to compound the problems.

While the rapid development in California had made it a State of the Union in 1851, there remained great areas of the West still unpopulated and unexplored. One of these was the central Rockies, a region steadfastly avoided by the trails and one that had been paid little attention since the demise of the fur trade in the 1840s. If the United States were to enjoy another El Dorado, 1858 was the time and Colorado Territory would be the ideal place. And there was little then, with all the problems, to hold men in the East.

The presence of gold in the Rockies had been first reported by French trappers a century earlier and confirmed later by American mountain men. The strike that would precipitate a gold rush, however, fell to a group of veteran Appalachian gold miners, dropouts from the California rush. They made the strike in the spring of 1858 near the present site of the city of Denver. When word reached the East, the press, promoters, and guidebook printers quickly distributed the typically exaggerated reports of the discovery to those who wanted, and now believed in, another major gold strike. By autumn of that year the nation was on the brink of its second rush, this one bearing the name of Pikes Peak, the dominant landmark of the discovery region.

In the spring of 1859 one hundred thousand Fifty-Niners armed with pans, shovels, and guidebooks hit the trails west for Colorado Territory. That year produced very little gold but a lot of disappointment. That turned first to disgust, then to bitterness, as a long line of wagons turned back, heading eastward, their drivers cursing the name of Pikes Peak. The more determined gold seekers remained and turned to the Rockies themselves to locate the sources of the disappointing discovery placers. Hundreds of small bands of prospectors, following a flurry of rumors as well as their own intuitions, ventured west into the rugged canyons. Within a matter of weeks, ten discoveries were made, each of which dwarfed the original 1858 strike. The streams of the Colorado Rockies were rich with gold. By fall of 1859 a string of mountain towns that are still alive today were founded. Idaho Springs, Fairplay, Central City, Breckenridge, Blackhawk, and Silver Plume are some of the towns started then. At its very worst moment the Pikes Peak gold rush had been redeemed, and the Territory of Colorado, like California, was now destined for greatly accelerated settlement and development.

ROCKY MOUNTAIN GOLD FEVER

In the pattern established in California, the first miners to reach the streams made their fortunes quickly. Those arriving late shoveled gravels for day's pay or settled into the developing placer mining industry.

Within a short time, the lode sources of the gold were

CHAPTER 5: AND STILL MORE GOLD

discovered and Colorado became a pioneer state in the development of modern hard-rock mining. The early 1860s saw the introduction of the first mechanical rock drills which, although nearly unworkable in comparison with today's drills, were a definite improvement over the ancient technique of hand drilling. The demonic roar of the drills shook the underground workings as steel bit into solid rock seeking elusive veins of gold-bearing quartz. The peak year, 1862, boasted eight tons of gold; by 1866 most of the excitement was over. But not before sixty tons of Colorado gold had pumped another $25 million into the United States economy.

A secondary effect of the Colorado rush was the dispersion of prospectors into other areas of the Rockies where many minor gold strikes were made. None proved the equal of Colorado's 1859 discoveries, but many gave birth overnight to "million-dollar" camps that thrived, however briefly, on both placer and lode gold production. They ranged from the deserts of Arizona and New Mexico, through the barren lands of Nevada, and into the mountains of Wyoming, Idaho, and Montana.

Even farther north, rich placers were discovered in the Canadian province of British Columbia. Hundreds of towns and the mines they served sprang into being, then died. Today these towns are bleached timbers and memories.

Prospectors explored every peak and canyon of the Rockies. In the process of looking for gold, significant discoveries of base metal desposits such as copper, lead, and zinc were made, but exploitation of these deposits was delayed somewhat because of the obsession with gold. Prospectors knew in their hearts that another major gold strike was imminent.

The practical mining codes adopted in the gold camps well served their purpose and were formalized by an Act of Congress in 1866, then amended and clarified in 1872. Never before had any nation been so free with the allocation and use of its lands. The legislation's basic purpose was to invite settlement, development, and use of the vast western lands. In that respect it succeeded fully, and no group put it to better use than those countless thousands of gold seekers.

The 1872 mining codes simply legalized what had been considered standard since the days of the California rush. Basically, any U.S. citizen was permitted to stake placer claims, fifteen hundred feet long and six hundred feet wide each. Clear title was granted after the claimee made five hundred dollars' worth of improvements and paid a modest five dollars per acre. Obtaining title, or patenting, was not necessary; as long as the annual improvement work was made and filed, or actual mining conducted, the claim remained effective. If the requirements were not met—say when the claim was abandoned—the land simply lapsed back into the public domain. The laws were far from perfect and suffered a good deal of abuse but, as intended, they stimulated mineral exploration and exploitation. The same laws remain in effect today.

GOLD IN THE BLACK HILLS

Colorado's gold rush was a decade past by the time the next big strike was in the making, this one in the southwest corner of Dakota Territory. Colonel George Custer's overland mapping expedition had taken him through the Black Hills where his mineral surveyors located rich placer deposits. When the word leaked out, it was received with jubilation and anticipation, but the inevitable rush was delayed for almost two long, tense years. A matter of some political importance took priority. It seemed that the Black Hills, the discovery site, had been ceded in perpetuity to the Sioux Indian Nation in an 1868 treaty. While the government pressured the Sioux for formal alteration of the agreement, the first impatient gold seekers risked their scalps (some gave their scalps) for a first crack at the placers. The Sioux bitterly resisted changing the treaty to allow waiting hordes of wild-eyed gold seekers to desecrate their lands. Nonetheless, it was the quest for gold that once again took preeminence. The Great White Father in Washington unilaterally lifted the government embargo on the sacred Black Hills of the Sioux.

The trails leading to the Black Hills in the spring of

CHAPTER 5: AND STILL MORE GOLD

1876 were churned to rivers of mud by the endless columns of racing horses, wagons, carts, stagecoaches, and men on foot. The streams were quickly staked, the forests of lodgepole pine leveled. And as the gold was taken out of the sluices, the city of Deadwood, one of the wildest boom towns of all, was built in a matter of days. As expected, the blatant breach of the treaty touched off a war with the Sioux. June 1876 proved to be a month of no little irony, for as the Black Hills gave up their gold to the shovels and sluices, the man responsible for the discovery, George Custer, was two hundred miles away to the northwest about to lead his men into the Little Bighorn.

THE HOMESTAKE LEAD

The first of the Black Hills gold came from placers, but by late summer, a hard-rock *lead,* or lode outcropping, was discovered near Deadwood. This site quickly passed into the hands of San Francisco financiers who provided capital for a major underground operation making full use of the new rock drills and modern dynamites. The town that grew to house the miners was named Lead, and the mine, the Homestake, went on to become one of the greatest gold mines in the world. By now Homestake gold has made the original Black Hills placer recoveries look like pocket change. From the deep shafts near Deadwood have come over fourteen hundred tons of gold. Still running at full production after a century, the mine yields about ten tons annually.

The development of the Homestake was a good indication of the way gold mining would go in the western United States. Although there was still one great gold rush left to act out, the richest deposits within reach of the individual already had been worked. Corporate operations took over the mining industry. Individuals were relegated to trying their luck on previously worked gravels, for the prospects of a new, large strike diminished as the Rockies became ever more thoroughly explored.

The corporate approach was encouraged by the new drills and dynamites, but even more so by advanced ore treatment and chemical extraction processes, the most important of which was *cyanide treatment*. Crushed ores treated with a solution of potassium cyanide would easily give up their gold in the form of soluble gold cyanide (one of gold's few compounds) from which the metal could be recovered inexpensively by chemical replacement with zinc or by electrolysis. The new recovery processes made it economically feasible to mine the lower grade ores theretofore ignored. More and more, even placer mining fell under corporate control. Low-grade gravels were attacked with enormous hydraulic pumps that literally moved mountains to reach small gold-bearing layers. The 1890s saw the introduction of the huge "gold boats," the floating hydraulic dredges that were first used in the gold fields of New Zealand. The dredges were essentially high-volume, self-feeding, movable sluices capable of chewing through thirty feet of overburden to recover gold from lower strata.

GOLD STRIKE AT CRIPPLE CREEK

The western frontier was rapidly becoming history when the mountains of Colorado provided one more golden surprise. In 1891 a down-and-out cowboy discovered rich gold ore in the shadow of Pikes Peak. Two years later hard-rock exploration had revealed the extent and richness of the deposits.

Cripple Creek and its booming satellite towns were born, and, by 1897, the area's 475 mines were producing twenty-six tons of gold, with a value of $10 million, annually. Production was to continue for forty years.

The ore at Cripple Creek was some of the most remarkable ever encountered. Occasionally miners would blast their way into a *vug*, or natural cavity, in the rock that would be lined with rare crystals of gold of extraordinary delicacy, size, and beauty. One such vug was nearly five feet in length with the entire interior surface lined with a crystalline gold latticework.

Such discoveries were rare. But because all gold miners, by the very nature of their job, were in close contact with the gold they were paid to mine, many were moved by the age-

CHAPTER 5: AND STILL MORE GOLD

old desire to keep a bit of that gold in their own pockets. *High grading,* as the controversial practice became known, meant different things to different people. To the miners, it was an honored and earned right, but to the mine owners, it was plain criminal theft. High grading in the placers was difficult, if not impossible, as the sluices were in sight of everyone.

"Keep shovelin' and don't bend down"—in other words, keep your fingers out of the riffles—was the order of the day for hired hands working the sluices.

Hard-rock mines were a far different story. They were dark, dirty, and incredibly dangerous, and owners rarely, if ever, ventured into the deep shafts and narrow drifts. Strict visual supervision of the miners was nearly impossible and totally impractical. So hard-rock miners found themselves in a tempting situation. Until well after the turn of the century, most miners received a daily wage of three or four dollars for a long shift of physically risky work. Yet a single ounce of gold, no larger in size than a quarter, was the equivalent of a full week's wage. Miners being human, not all the mined gold went into the ore cars—some found its way into pockets and lunch pails. Personal searches of miners coming off shift sought to discourage the pilfering, but most miners took this as a sporting challenge and devised ingenious ways of smuggling out the gold. The smugglers held no bodily orifice sacred. Criminal charges of high grading rarely brought convictions since the jury was packed with miners, and every one of them knew he could be sitting in the same dock a week later. High grading reached epidemic proportions, especially in those mines where rich veins would yield twenty or thirty ounces of gold per ton of ore. The actual amount of high-graded gold will never be known, but some "poor" miners had already "earned" many times their company wages by payday.

THE GOLDEN NORTH

Another great gold rush was in the making. In the far North prospectors had tramped the Yukon River drainage for a decade, always finding just enough gold to keep them interested. Finally, in August 1896, rich placers were discovered in Canadian territory near the confluence of the Yukon and Klondike rivers. The individuals to profit most were those already in neighboring Alaska who could reach the strike quickly. The best gravels glittered with gold, and miners could wash over one ounce from a single pan. By summer and fall of 1896 miners on the better claims were taking five hundred to one thousand dollars in gold from their sluices every day.

When word of the Klondike strike reached the United States in January 1897, the promise of gold galvanized men then even as it had in 1848. By spring the lead columns of what would become a full-scale stampede headed north by two long, arduous routes. One was by river steamer up the Yukon River in a thirty-five-hundred-mile journey. The other involved an overland trek from the southeast Alaskan coast to the upper Yukon Drainage, a shorter but much more difficult route. The first successful miners, satisfied with their gold and glad to leave the rigors of the North, returned to the United States in the fall of 1897. When they stepped onto the Seattle and San Francisco docks with their strongboxes of Klondike gold they had washed out in a few months, even the skeptics booked passage north. Steamship companies reorganized schedules and leased ancient, unseaworthy hulks to handle their booming trade. By 1899 over one hundred thousand gold seekers had rushed for the Klondike.

The year 1900 alone saw twenty-two tons of gold worth $10 million taken from the Klondike gravels. With no streams remaining unstaked, hordes of prospectors fanned out through the vast Alaskan interior. Some struck it rich and triggered smaller local rushes to such places as Nome, Fairbanks, and a hundred lesser camps. Alaska only four decades earlier had been referred to as "Seward's Folly" by the detractors of the man who urged its purchase from Russia for the then-considerable sum of $7,200,000. Seward's Folly has since produced over fifteen hundred tons of gold with a value exceeding $1 billion. In the year 1900 gold was the only force on earth capable of luring substantial numbers

CHAPTER 5; AND STILL MORE GOLD

of men to the remote Alaskan interior to perform the settlement and exploration that eventually led to the state's development.

END OF AN ERA

By 1901 the Klondike rush was over, and the miners settled into the slow, methodical washing of lower grade gravels. The wild American frontier adventure with gold was over, but what an adventure it had been. The gold taken from the North American Rockies from 1848 to 1900 was nearly four thousand tons, several times all that the Spanish had gained through three centuries of conquest and exploitation. The fifty-year sequence of western gold rushes put more gold into circulation than the world had mined in its entire history.

Gold had directed the politics of a continent, led to the opening of huge tracts of previously unexplored land, shifted the economic balance of the world, become the basis of the economy and currency of the United States, and spurred the development of mechanical and chemical innovations that would aid the mineral industry around the world. Gold had influenced our music, literature, and speech, and some of the most colorful chapters of American history were written in gold ink. Probably as many as two million individuals had hunted for gold. A few were made millionaires, a few more were able to retire to the life of their choosing, and even more, after meeting the challenge, had come out a bit ahead. The great majority received their reward not in golden wealth, but in the form of adventure and experience that both altered and enriched their lives. Behind them they left the mountains with uncountable placer and hard-rock workings, some of which had indeed yielded fortunes in gold, and many more that had granted only disappointment. Like the Spanish before them, these gold hunters had taken a great deal of gold from the ground but left far more behind for those who would someday follow them. However, by the outset of this century the era of the individual gold seeker was temporarily at an end.

6
OUT OF SIGHT, OUT OF MIND

At the time the United States won independence, the traditional value of gold was about sixteen times greater than that of silver. The silver dollar coin—a direct descendant of the Spanish eight-reale silver piece commonly known the world over as the *piece of eight*—contained about one Troy ounce of silver. The value assigned to one Troy ounce of fine gold was twenty silver dollars (the variance from the traditional ratio due to the minting costs). This equivalence was formalized in 1837 when the fixed value of one Troy ounce of fine gold was declared to be $20.67 in United States money. That was the theory anyway.

In practice, a miner did not receive the full value for his gold, but a lesser amount determined by its particular purity and any associated refining fees. And since the United States had decreed, at the same time the price was fixed at $20.67, that all mined gold be sold to the government, a miner had no legal choice but to take what was offered. Most California placer gold, for example, was purchased for $16 per Troy ounce, as it was approximately 80 percent pure. But the formally fixed price of gold remained unchanged well into the 1900s.

MONEY IS A FUNNY BUSINESS

In 1879, at a time when the nation was still suffering from the financial drain created by the Civil War, a confusing depreciating paper money standard was employed for the first time. Theoretically, this meant rejection of the "standard" silver dollar. Technically, the nation remained on its bimetallic standard, but, in reality, was already informally on the gold standard.

In 1900 the Gold Standard Act formally and legally placed the United States on the gold standard to join most other Atlantic trade nations. The act declared that the gold dollar "shall be the standard unit of value, and all forms of money issued or coined by the United States shall be maintained at a parity of value with this standard . . . ."

Under this system, the standard unit of value was the gold dollar which was composed of exactly one Troy ounce of fine gold which in turn could be coined into $20.67 in United States money. Furthermore, under free coinage, a two-way situation existed where anyone taking one ounce of gold bullion to a U.S. mint to be coined would receive $20.67 (minus assaying and refining charges, if any), and a person melting down $20.67 worth of United States gold coins would end up with a mass of gold weighing exactly one Troy ounce. The old bimetallic standard was forever put to rest, the price of gold was formally fixed, and silver was allowed to seek its free price through normal supply and demand.

The price of gold had remained fixed for nearly a century at $20.67 per ounce, a level stubbornly maintained in the face of constant inflation that characterized a greatly expanding economy. Remaining artificially controlled both domestically and internationally, gold became worth less each and every year in terms of the hard reality of purchasing power, the ultimate measure of real value. The gold mining industry was the first to suffer, as its profit margin declined steadily. Full production was no longer the word, for concern now lay in mining only the deposits that could be worked most cheaply.

CHAPTER 6: OUT OF SIGHT, OUT OF MIND

In the West, along the thousands of miles of old placers, a few grizzled old-timers clung stubbornly to their claims, fighting the inevitable. Every year the sixteen dollars they received for an ounce of their gold bought less, and their costs always increased. During the 1920s in many placers all over the West, the cost of washing one ounce of gold from the gravels became more than sixteen dollars, and the sluices were left to rot and silt over. Where nature could never tarnish the luster of gold, the artificial controls of big government succeeded.

With the exception of a twenty-month period during World War I when it was necessary to restrict gold exports, the gold standard continued in operation until 1933 at which time the system failed utterly and completely.

In the end it was the common man's faith in gold, and nothing else, that ruined it. Specifically, in the United States, it was a lack of confidence on the part of the common man in paper currency that caused a run on the banks where endless lines of people waited to convert their paper into gold. In a few short weeks the run on the banks had depleted the national gold reserves by a staggering 75 percent.

What followed was a series of acts of Congress and government moves that revoked the gold standard, making it illegal to export or even own gold, with the exception of jewelry or numismatic pieces. Gold coins were declared no longer legal tender and all outstanding coins were ordered to be turned in to the government in exchange for paper currency. On 1 January 1935 a formal devaluation of the dollar reestablished the price of gold at thirty-five dollars per ounce. The price of gold is still controlled.

THERE'S LIFE IN THE OLD MINES YET

In only one year following the devaluation of the dollar and the corresponding rise in gold, domestic gold production in the United States doubled. None of the gold deposits in the country had been mined out, as many had believed. After one hundred years of price control on gold, they had simply not been worth mining. Corporate mining operations returned to full production, moving more tons of ore than ever before. And all through the Rockies, in the placers and hard-rock workings long dormant, the sounds of the shovels and drills were heard once again. The rewards of the quest were once again worth the effort.

Both individuals and companies geared up over the next few years, only to be frustrated once again in 1942 when the War Production Board issued Order L-208. All gold production was immediately terminated so that the total mining effort could be directed solely towards the production of the vital base metals. A great number of gold operations, especially the hard-rock mines which required pumping and constant maintenance, shut down never to reopen.

TIME FOR SOME MONEY MAGIC

Once again the price of gold remained fixed by governments while inflation, steeper than ever before, eroded away gold's real purchasing power. After World War II gold was valued not enough to lure men back to the back-breaking placers and small hard-rock mines. Corporate mining ventures had little choice but to maintain limited production so as to at least break even while maintaining underground workings.

Most of the gold production in the United States now came not from primary gold mines, but rather as a by-product from the base metal mines. Maintaining an approximate annual production of one hundred twenty-five tons, the United States fell to fourth in world production, far behind the Republic of South Africa, the Soviet Union, and Canada. In 1970 the U.S. Bureau of Mines stated that only five true gold mines—four hard-rock and one placer—still remained active out of the twenty-five previous top producers.

Since the demise of the gold standard in 1933, gold in the United States had become a matter of "out of sight, out of mind." As inflation continued and the demand for gold outstripped world production, gold, in the 1960s, had become a notoriously undervalued commodity. Paper currency enjoyed no worldwide confidence at all, and international de-

CHAPTER 6: OUT OF SIGHT, OUT OF MIND

mand as well as outright hoarding made the United States subject to a foreign gold drain.

The world once again made its choice. It favored neither the political and economic systems that had been established, nor the various forms of money that had been created. Instead, it shouted once agian for gold. The world was ready, in fact, overdue, for a major international monetary adjustment. It was time to set things right.

7
GOLD FEVER AGAIN

A modern El Dorado was waiting in the wings. As early as 1960 speculation surfaced that the United States would attempt to deter its overseas gold drain by devaluing its dollar against gold. In anticipation of this gambit, the price of gold on the international market climbed to forty dollars per ounce. In a stubborn effort to hold the price at the fixed thirty-five-dollar level, great quantities of gold bullion were dumped on the market. Some success was realized: the price line was generally held throughout the 1960s, although at enormous expense to the free world gold reserves.

BREAKING POINT REACHED

The strain of holding the price had reached the breaking point by 1968 when Great Britain devalued the pound sterling against gold. The actual extent of the demand for gold—a demand created by the war in Vietnam—became apparent. Throughout Southeast Asia the chaotic political and military situations made everything about the future perilously uncertain, with the single exception of gold. Any Southeast Asian affected by the war (meaning everyone) who possessed any assets to speak of converted them to gold. It mattered not, from the standpoint of personal economics, which power controlled which nation; there was no power on earth that would make gold worth one *piastre* less while everything else of material value in Southeast Asia was vulnerable to destruction, devaluation, or failure.

The international demand for gold was further increased when fabrication consumption reached new levels. By 1970 over fourteen hundred tons of gold were being fabricated annually. Most of that, one thousand tons, went into jewelry which was in greater demand than ever, not only for its traditional ornamental value, but for its new investment and speculative value. Coins, medals, and medallions accounted for another one hundred twenty tons. A relatively new and quite substantial utilitarian demand saw two hundred tons going into aerospace, electronic, and industrial applications. In 1970 about thirteen hundred tons were mined, one hundred tons less than the fabrication demand.

In 1968 an awkward two-tier system was employed in international markets. Central banks traded gold at the fixed price of thirty-five dollars per ounce, while other interests (foreign) could buy and sell at a free-fluctuating free market price. This was not yet a true free market, for the fixed central bank price discouraged outright speculation. Still, by 1972, the "free" price had drifted up to nearly seventy dollars, double that fixed by the banks.

SUPPLY AND DEMAND HELP ADJUST PRICE

In 1973 the United States devalued its dollar, driving the free price to nearly ninety dollars per ounce. At this time most nations abandoned the futile and costly attempts to hold the line on gold, allowing the price to seek its own honest level as dictated by open supply and demand.

And then on 31 December 1974 the United States legalized individual ownership of gold for the first time in forty years. Investors, hoarders, speculators, collectors, as well as the merely curious, from grey-haired grandmothers to wealthy corporate interests, turned to gold as never before.

CHAPTER 7: GOLD FEVER AGAIN

Their reaction demonstrated confidence in international politics and economic structure. That confidence proved minimal, for the free interaction of supply and demand soon drove the price of gold beyond even the wildest predictions. That price, controlled for a century and a half at levels of twenty dollars, then thirty-five dollars, skyrocketed. At its peak, the price of one Troy ounce of fine gold neared the incredible price of nine hundred dollars.

This dramatic adjustment in the price of gold also served to greatly enhance the metal's qualities of liquidity and portability. Accompanying the free marketing of gold was the massive manufacture and sale of bullion and bullion coins for general distribution and circulation. Today there is almost no place in the free world where these items cannot be converted immediately into paper currency or directly to material goods. Even in the communist world, where the private ownership of gold remains a crime, the black marketing of bullion and bullion coins at prices substantially in excess of the free world price is common.

The greatest boost to the practical liquidity of gold has come from the universal acceptance of the bullion coins, of which the South African Krugerrand is the most common. The words *South Africa* that appear on the coin are meaningless, for the acceptance of gold goes beyond the allegiance to any nation. Nowhere on the Krugerrand is there the familiar and traditional statement of fixed value. The only words of true importance appear on the reverse. They are *1 Oz. Fine Gold,* the universal statement of unquestioned and immediately convertible value.

CARRY A LITTLE, CARRY A LOT

Gold is portable, and with the spectacular rise in price, you don't have to carry much to be carrying a lot. Compared with other materials which have been used in the past for money, including cattle and land as extreme examples, gold is about the ideal medium of wealth or exchange. A single pound of gold, as of this writing, has a value of over seventy-five hundred dollars, yet occupies less space than that taken up by twelve neatly stacked U.S. half-dollar coins. A single bank roll of forty half-dollar coins, if they were fine gold, would be worth over twenty thousand dollars. At current gold prices, the proverbial million in gold would take the space of only about fifty rolls. Be sure your cash drawer is sturdy, for those rolls will weigh almost one hundred fifty pounds.

FUTURE PROSPECTS ARE PROMISING

Turning toward the future, the expected demand seems certain to exceed any foreseeable contributions to supply. Gold production in 1980 was among the highest in history with a total of nearly fifteen hundred tons but still below rising fabrication demands. Fully half of the world's supply is provided by one nation—the Union of South Africa—and there are serious doubts as to just how much longer the great hard-rock mines of the Rand will be able to maintain output. Ranking second in current gold production is the Soviet Union, accounting for about one-quarter of total production. The extent of its ore reserves, estimates of future production, and just what will be done with current bullion reserves in Russia are all very uncertain issues.

Free world mining interests are now showing great interest in ore reserves that were considered too low grade to mine only ten years ago. Large mines are showing excellent profits now, working ore which averages only one-third ounce of gold per ton. When all factors have been considered, it is a good bet that the price of gold will achieve new highs rather than sink into any prolonged and significant decline.

There is little disagreement among market analysts that gold, over the long term, will surpass the one-thousand-dollar-per-ounce level. The only point they disagree upon is how soon this will happen.

It has been a long, long time since man first dug through the gravels in search of gold. Through the ages the yellow

CHAPTER 7: GOLD FEVER AGAIN

metal has never diminished, but only grown, in value, and men still go in search of it. In the Americas that quest spans nearly five hundred years of recorded history but had its beginning several milleniums before that in early Indian times. Again, today the quest for the gold of the Americas is very much alive, more so than at any time this century.

Backed by an extraordinary economic advantage and armed with the tools of modern technology, these are the most effective and efficient gold seekers ever. And the time has never been better to seek gold from the rock, the sea, and the graves of the Americas.

The accepted historical reason for the end of a gold rush or the closure of a mine has always been that the deposits were exhausted. This is a classic oversimplification since no gold deposit has ever been completely mined out.

II
FROM THE ROCK

8
MOUNTAINS OF GOLD

Those first pieces of yellow gold that early man plucked from the gravels some twenty thousand years ago were already the product of an infinitely more ancient geological process. Compared with the origins of most other mineral deposits, it was a relatively simple process. The earth, in its original form some 4 to 6 billion years ago, was a gradually cooling molten orb with all ninety-two natural elements present in either their liquid or gaseous states. As the cooling occurred, the elements and chemical combinations of elements became roughly separated according to their various densities. Those that would remain gases at the temperatures and pressures later considered "normal" remained at the surface to form the developing atmosphere. The heavier elements and compounds, including gold, sought the lower levels of the molten strata which, as cooling progressed, solidifed to form the crust of the earth.

FORMATION OF THE MOUNTAINS

The original surface crustal formation was composed primarily of the light elements and compounds, such as silicon, carbon, and the alkali metals. It immediately became subject to enormous stresses caused both by its own continuing contraction as well as the pressures generated by containment of the still-molten lower strata. The result was the formation of a variety of surface features, including mountains much higher and more massive than those in existence today.

This formative process was sustained over a billion years, causing still greater pressures within the molten lower strata as further cooling took place. Quantities of the still-molten material containing the heavier metals and minerals were eventually forced upwards into the cracks and crevices of the still-buckling crust. Some 60 million years ago, a time span considered relatively recent by geological standards, a series of major upheavals created a nine-thousand-mile-long complex chain of ridges and mountains stretching from the lands we know today as Alaska to Chile.

These were the mountains we now refer to as the Rockies, and they were built from the already eroded base of an earlier range previously mineralized by the rising molten materials. The creation of the "new" Rockies provided many cracks and fissures in the crust where still more of the molten minerals could be forced upward.

The molten material that carried the metals upward was composed mainly of magma and other liquified base rocks. Upon final cooling of the crust, these molten materials solidified to form mineral veins and bodies called intrusions, many of which were immediately attacked by the action of freezing and thawing water, atmosphere, and chemical alteration. As the upheavals diminished and the crust stabilized, the newly formed surface features and their mineralized areas were broken down, eroded, and altered both chemically and physically.

Among the heavy metals which, from time to time, were carried to the surface, was the metal gold. While the gold appeared in a variety of minerals, most often it was jets of molten quartz that bore it upward. Many other metals formed complex mineral compounds upon cooling, but the basic inertness of gold insured that most of it would remain in its nearly pure state, looking precisely as it does today—a lustrous, sun-yellow metal. The gold remained free of chemical alteration but was broken down with the other rocks into bits and pieces. Along with the huge masses of crumbled rock that formed the gravels of alluvial deposits, the crumbled gold was also carried away to be redeposited elsewhere. The gold that remained *in situ,* in its original place within solid rock, would become known as *lode gold,* and that which was broken down and redeposited elsewhere would be called *placer gold.* Gold, appearing exactly as it does today, was waiting to be discovered in stream gravels and in rock outcroppings millions of years before earliest man ever walked the earth.

THE REAL GOLDEN RULE

"Gold is where you find it," proclaimed gold rushers in the West. There seemed to be no set of rules dictating where gold would occur and could therefore be found. Gold had been found in streams narrow enough to step over, in broad, ancient river banks beneath fifty feet of worthless overburden, in dry placer deposits, and in other placers hundreds of miles from the nearest known lode source. It had been discovered in highly mineralized regions alone and in conjunction with other metals and ores, and also in areas that revealed no other mineralization other than the gold itself. It occurred in tantalizing trace amounts as well as in glittering pockets the early miners called "glory holes." Men had found it as dust as fine as flour, in weighty nuggets, as delicate wires weaving through quartz, and as complex ores bearing no visible indication of the metal they contained. Furthermore, gold had been found beneath the frozen tundra of Alaska, in the scorching deserts of the Southwest, among the highest peaks of the Rockies, and also in the steamy jungles of the tropics.

Gold being "where you find it" had an element of truth, but not in the broadest sense, for it was valid only within the limits of a known auriferous, or gold-bearing, region. During the 1800s the location and extent of these gold-bearing regions were still being determined by prospecting. In that era gold might very well be discovered where it was previously unknown, simply because the particular area had never been prospected by knowledgeable men.

Today the auriferous regions of North America are thoroughly and accurately delineated. Within those limits virtually every single stream and land feature has been prospected for gold at one time or another. Doubtlessly there remain a number of remote locales where only a hunter or someone not concerned with gold has visited, but they are rare and becoming more so each year. One's chances of discovering a gold-bearing stream unknown to previous prospectors are extremely poor, and the next unknown major deposit will be discovered by corporate prospectors, probably by deep core drilling techniques.

But today's gold seeker may still find a good deal of truth in the old saying, if a qualification is tacked on. Gold is indeed where you find it—*as long as you look where others have found it before you.*

LIKELY SPOTS FOR PROSPECTING

It should be noted that this does not at all hold true for South America where there are still enormous tracts of land never prospected. Unlike North America, the auriferous regions of South America have still not been entirely determined. As this book is being written, extremely rich placer strikes are being made in Brazil in areas that have never produced gold before.

In North and South America the most prolifically auriferous regions coincide with those areas that underwent the most severe formative buckling and upheaval and that are in or adjacent to mountain ranges. The Appalachians were formed in an early geological era and were originally about ten times their present size. Over the ages erosion has greatly reduced them in mass, with the alluvia forming the broad plains both east and west of the center of the range.

The Appalachian gold found in the early 1800s came entirely from placers, as the original lode sources have long been

CHAPTER 8: MOUNTAINS OF GOLD

eroded away, a point which dispels the myth that once someone has located a gold placer, all he must do is "walk upstream until he finds the lode source." There are still a few small placer operations in the southern Appalachians, and amateur gold panning is popular, but they cannot be considered a potential site for the serious gold seeker.

For these sites, great in both number and potential, the gold seeker must turn to the Rockies, the general name for the geological backbone of the Americas, that is one of the most continuously auriferous regions on earth. Throughout pre-history and recorded history, the Rockies, including their southern range, the Andes, have yielded something on the order of 800 million ounces—over thirty thousand tons—of gold, about one-third of the world's total production. That enormous production has led to another popular misconception about gold, namely that the deposits have been mined out. The accepted historical reason for the end of a gold rush or the closure of a mine has always been that the deposits were exhausted. This is a classic oversimplification since no gold deposit has ever been completely mined out.

Mines and even entire regions were abandoned by the thousands for a variety of reasons, not one of which has to do with the last bit of gold being extracted. The single overriding reason had to do with basic economics in an era when gold was worth far less than it is today. Geologists estimate that for every ton of gold mined, more than half that much still waits in the rock and is recoverable today. And most of it is in the placer and lode deposits of the Rockies that have already been discovered and partially worked.

9
PLACER GOLD

The first gold collected by man came from a placer deposit, and the first "mining" ever performed was the bit-by-bit hand gathering of small nuggets. Soon after that man arrived at the idea of washing the heavy gold from the much lighter gravels. And with that, placer mining had already developed, a technique so basic that it is still employed today with very little change. Placer mining remains the simplest and least expensive form of mining, for it is nothing more than an improvement upon the natural process that deposited the gold in the first place.

ORIGIN OF PLACER DEPOSITS

That natural geological process took place after the lode sources of gold had been deteriorated and broken down by water and erosion, with both the gold and common rock becoming part of the alluvial mix. In the course of movement to lower elevation, mostly assisted by water flow, the particles of gold, being far heavier than the accompanying rock, worked their way by gravity to the lowest levels. As the gravels sorted and resorted themselves, the gold formed concentrations at the lowest possible levels of the loose alluvia. In streams and rivers this occurred on or very near bedrock, the in-situ rock over which the water and alluvia moved. The degree and manner of concentration of gold was dependent upon an infinite number of factors which is reflected in the wide variety of types of placers found today. The mechanics of placer formation in nature are exactly the same as those allowing gold to be collected in pans and sluices.

SIMPLE TOOLS — GREAT REWARDS

By the early 1800s the basic tools of placer mining, the gold pan and the sluice, were already in use in the Appalachians and in that form would serve the gold seekers of the West from California to the Klondike in the coming century. The great appeal of those gold rushes stemmed not only from their historic timeliness and the ancient dream of a golden fortune, but also from the fact that the great strikes were made on placer deposits. Placer mining, it seems, is made for anybody, who needs to provide only courage and determination. Placer mining requires no geological or engineering knowledge whatsoever, simply an understanding of the basics of placer formation. The cost of outfitting with the essential tools is next to nothing: a washbowl is a gold pan and every tree is a potential sluice. Humble tools wash gold from the gravels.

To understand placer mining, a man has only to understand the sluice and the gold pan. The first gold pans employed thousands of years ago were ceramic or wooden vessels, and the Spanish batea was a somewhat improved version of an already ancient tool. Metal pans were manufactured in the United States in the 1820s in quantity. Since that time volumes have been written on the proper use of the gold pan, mostly in recent years and aimed at the "amateur" gold panner. In terms of technique it is very hard, almost impossible, to go wrong, even though a survey among one hundred different panners would probably reveal one hundred subtle variations of exactly the same thing.

HOW TO PAN FOR GOLD

There is but one all-important point to remember, and that is to be absolutely certain that the material in the pan has been thoroughly washed and retains none of the claylike cohesion and consistency common to most gravels. This includes washing material from the surface of each and every large stone that finds its way into the pan. When the mixture of gravel and water has achieved complete fluidity, five or ten seconds of vigorous motion—shaking, swirling, or anything similar—will make any gold drop to the bottom of the pan. The uppermost layers of the material may then be discarded by pouring or washing off while watching carefully for a possible nugget which, because of an unusually flat shape, may have been carried upward. The functional shape of the pan, together with a bit of common sense will assure that the heaviest materials present will remain securely at the bottom of the pan. Depending upon the cohesiveness of the gravels, washing a pan should never take longer than just a few minutes.

What now remains in the pan, the concentrate, must be treated more carefully. The concentrate, which should not amount to more than a few ounces, will be very dark, almost black, in color. These are the *black sands,* the significance of which has been widely misunderstood. Black sands are composed primarily of crystalline grains of heavy minerals, most commonly magnetite and ilmenite, the magnetic chemical salts of iron which are among the most common and widely dispersed minerals on earth. Their presence in no way assures or even indicates the presence of gold. However, *if* gold is present, it will be found in the black sand concentrate, or in the lowest strata of the stream bed itself which is naturally rich with the heavy black sands.

Because of relatively close specific gravities, black sands and particles of gold will prove increasingly difficult to separate. The first step from this point is to dry the concentrate to a powder, thus allowing most of the black sands to be easily separated magnetically. If the gold present is of the "flour" variety and virtually impossible to separate cleanly and without loss, a simple mercury amalgamation will suffice; but if one does not wish to attempt further separation, any refiner's purchasing agent will gladly accept placer concentrates containing only a minimum of 10 percent gold.

HOW TO WORK A SLUICE

The basic tool for production placer mining is the sluice box, nothing more than a modified and controlled stream bed in miniature. A sluice is simply a trough with a series of riffles, or ridges, on the floor serving to hold gravel and rock which in turn act as a very effective trap for gold particles. Sluices range in length from two to ten feet, but many miners use eight-foot units, the length of a standard sheet of plywood.

Once set in the stream bed so that water is coursing through it, gravel is shoveled or otherwise introduced into the head of the sluice. From that point, successful operation is largely a matter of common sense. Water speed must not be so fast as to rush gravels through without time for separation and settling, nor so slow as to permit clogging with worthless common gravels. The remaining consideration is to be sure the gravels are broken down completely by the turbulent action of the water before they are washed clear of the sluice. If the gravels are extremely cohesive, a prewashing may be necessary to insure complete separation.

Two men with nothing but shovels can easily put as much as ten tons of gravel through a sluice in a single day. At the end they will have only a few panfuls of concentrate from the sluice which may be further refined by careful panning. Proper use of the pan and the sluice, together with a working knowledge of placer formation, is all that is required to make one a competent placer miner.

MOVE GRAVEL TO FIND GOLD

Prior to the Spanish exploitation and the nineteenth century gold rushes, most of the streams in the Rockies were textbook examples of orderly, natural alluvial sorting, separation, and deposition. In the broader, more ancient streams and rivers, some degree of rechanneling was normal and left behind old formations of banks and shelves of gold-bearing gravels while another placer was started in the most recent watercourse.

In the younger streams, particularly those flowing over solid rock in the higher elevations of the mountains, natural rechanneling was impossible, and many concentrations of gold became highly enriched over the ages. These were the deposits that the Indians mined, that awed the Spanish, and that fostered the gold rushes. Such deposits are extremely rare today, both because of removal through mining as well as the effect of the mining which altered the deposits and stream formations in every way conceivable.

In their quest for the gold, miners dammed streams, diverted channels of water to erode away banks, hydraulicked whole hills into mud and gravel, exposed bedrock, bulldozed massive heaps of gravel aside, and blasted away rock formations that got in their way. Fifty years of gold rushes had altered the streams more than nature had in a million years. There are virtually no gold-bearing streams in the Rockies not affected by mining.

GOLD IS GOLDER ON THE OTHER SIDE

While all streams have been altered to some degree, it in no way reduces a miner's chances of successfully working placer deposits today. Thanks in part to the attitudes and priorities of the early miners, very little of the actual gold present in the placers was actually recovered. In the great excitement of the gold rushes, few miners would content themselves with a modestly but steadily producing operation while a steady flow of wild stories and rumors of glory holes and bonanza strikes just "over the hill" reached their ears every day.

Gold rushes were founded not upon the idea of a profit or a livelihood from the placers, but upon nothing less than the dream of a fortune. For most of the footloose miners, the thought of someone else discovering the next El Dorado while they worked a lesser stream was not easy to accept. Accordingly, most miners took only the richest and the easiest gravels, leaving the rest to go in search of better. Entire producing gold camps were sometimes closed down in a matter of days upon receiving word of the big strike somewhere else. Even the late arrivals were a bit hesitant to invest their time and sweat on a previously worked placer for the same reason. The fact that a placer had been abandoned meant that better gravels could be found on other streams. Miners might not return to their workings the next season for any number of reasons, knowing well that they had mined only a small part of their gold.

The vast majority of all final abandonments were simple matters of economics. At only sixteen dollars per ounce, even some of the better gravels simply were not worth the effort.

The early miners were ingenious when it came to devising ways to mine the gravels they wanted to, within the limits of their crude techniques of damming, diverting, and digging. Hydraulicking was the first and only example of modern technology being applied to placer mining until recent decades.

NOT MINED OUT

Throughout the gold rushes the recovery efficiency—that is, the percentage of all gold present in washed gravels actually recovered—was not at all great, amounting in some cases to only 50 percent. On every gold-bearing stream, there are banks and shelves that have never been mined and many deposits at the bottom of active channels beyond the reach of early miners. Geologists and mining authorities estimate that as much as *three-quarters* of all the placer gold ever deposited by nature remains in the stream gravels.

10
LODE GOLD

Hard-rock mining, the underground extraction of in-situ ore, has always been and will always be a difficult and costly process. Both the Egyptians and Spaniards took gold from rich hard-rock mines, but only because of the availability of unlimited, expendable slave labor. Except in the case of extremely rich ore, early hard-rock mining was economically impractical until the introduction of modern drills and dynamites. After that, hard-rock mining became commonplace all over the Rockies. Today there are an estimated twelve-thousand-plus abandoned hard-rock gold workings from Alaska to the southern Andes. A few are primitive Indian workings, others are the remains of Spanish mines in Central and South America, and most are the tunnels and shafts made by the frontier miners after the 1870s.

A JOB FOR A FEW GOOD MEN

The early hard-rocker, as well as his contemporary counterpart, are a breed apart from the placer miner. Any man with a few tools and a lot of hope could become a placer miner. But a hard-rock miner needed a working knowledge of the mechanics of underground mining, at least superficial mineralogical knowledge, and the ability to accept working conditions that were dark, dirty, and often terribly dangerous. Also needed was a substantial amount of capital. Hard-rock mining required a greater investment not only in money, but also in effort and time, and was based on more complex economics than was placer mining. During the long decades when gold was valued at twenty dollars an ounce, a miner would receive only about eight to ten dollars for every ounce of gold his ore yielded. During the late 1800s small mine operations were reluctant to touch ore that did not assay out to five ounces or more of gold per ton. It just wasn't worth the effort. A "bonanza" hard-rock mine at this time was one that produced ore from relatively shallow deposits containing as much as twenty ounces per ton. It is interesting to consider that most of the ore produced by the remaining primary gold mines today contains far less than one ounce per ton.

Hard-rock mines were rarely abandoned on whimsy. Too much time and energy was involved. Unfortunately for the men who worked them, most small hard-rock mines were hit-or-miss affairs, especially in the days before core drill sampling was possible. The small workings simply followed gold-bearing veins which often "pinched out." Whether the vein would "come back in" was a matter of the purest speculation.

Many early mines were closed when they were within drilling distance of veins.

Hard-rock abandonments were caused by many factors, one of which was the secrecy under which they were usually worked. Early hard-rock miners and prospectors were not known for their trusting or gregarious natures. Without estates, the death of the claim holder was often followed by the mine reverting back to the public domain. Many small mines active in the 1930s to the 1950s were forced to shut down when government mine safety laws upped operating costs.

In the end it was economics that closed most of the small hard-rock mines. Mining costs rose steadily every year, but by the early 1900s the fixed price of gold had made unprofitable even the ore assaying out to several ounces of gold per ton.

Shipping was another problem. Ore was technically worthless until it was delivered to a milling and smelting facility. The burden of shipping fell on the miner and, especially in the more remote and inaccessible mines, was a major expense. Only the higher grade ores warranted the expense of slow and laborious hauling over rutted wagon roads to a smelter. Much of the lower grade ore was stockpiled next to the waste-rock dumps at the mine to await an improvement in shipping or access, or a rise in the price of gold. Heaps of ore containing several ounces of gold per ton were frequently left when the mine was abandoned. While that abandoned ore was deemed low grade in 1920, it was very much high grade in 1980.

Almost all of the thousands of abandoned hard-rock workings in the Rockies are surprisingly small. They were built with just enough room to allow movement of miners and their equipment. No rock was moved unnecessarily. Entrance was provided through a portal to a horizontal tunnel, or through a vertical shaft descending to horizontal drifts at lower levels. Any shaft required a *headframe,* a surface structure making possible the raising and lowering of men and supplies and the removal of waste rock and ore. At the time of abandonment, some were dynamited shut, while others were left open.

All abandoned mines have one point in common today: they are extremely dangerous. Timbers have rotted away and fallen out of place, and the rock itself, through exposure to air and water, has cracked and weakened. Every year rockfalls and cave-ins claim more than a few lives.

PROFIT IN THE HARD-ROCK BUSINESS

The rise in the price of gold has greatly increased interest in hard-rock development, but the decision to go ahead with a hard-rock venture should be based only upon expert professional evaluation. Even the smallest mine will require more capital than one man or a small partnership may be willing to risk. Delineation of veins or ore bodies is an expensive prerequisite to development of a new mine or restoration of an abandoned working. The costs of compressors, drills, timbers, pumps, explosives, and other equipment and supplies will drain quickly even a large grubstake. Still further costs will be found in meeting safety, health, and environmental regulations as well as in preparing for the state and federal mandatory inspections. Even with the rise in the price of gold, most small hard-rock mines will be a borderline economic situation.

Even so, independent assayers are performing more assays for gold on hard-rock samples submitted by individual prospectors than ever before. Most of these prospectors are not themselves developing mines. What they *are* doing is searching the mountains and deserts for deposits or veins that may be claimed and then sold to the larger mining companies for either immediate development, further exploration, or speculative holding. Several prospectors have recently done very well in this area, selling claims for over one hundred thousand in cash and a small percentage of any future production.

FIND IT NOW WHERE OTHERS LEFT IT

Early prospectors missed very little in the way of visible gold. The classic quartz-gold outcroppings have long been hammered into oblivion or systematically mined. Most of today's hard-rockers are looking where others have found gold before them—in or near the thousands of abandoned workings.

The use of new electronic instruments makes each and every one of those old mines of the Rockies in North and South America a potential source of immediately recoverable gold.

II
New Technology

Until recently gold seekers used the same tools and techniques that have served for centuries. The first real benefits of an emerging modern technology were enjoyed not by the individual, but by the corporations that profited greatly through use of the new drills, dynamites, and chemical ore-treatment processes. It has only been since the 1950s that technology came to the aid of the individual. Since that time new tools and techniques have provided today's gold seekers with advantages his predecessors could only dream of.

TECHNOLOGICAL WIZARDRY

Man has fantasized for thousands of years about a device capable of sensing or detecting the presence of *unseen* gold. Just as some alchemists put their imaginative efforts into the creation of gold, others tried to invent instruments that could "find" gold. Most attempts concentrated upon finding the right combination of a directional indicator, such as a wand, and an input of psychic energy that would literally point at hidden gold. The attempts were original and colorful, but success was not realized until modern times.

The first breakthrough was accidental. In experimentation with radio direction-finding equipment about 1930, it was noticed that distant steel objects caused disturbances on receiving instruments. World War II stimulated technological research and development, including that of the new metal detectors. The immediate result was the development of unwieldy land-mine detectors employed in field use to detect the presence of land mines planted by the thousands.

Operation of the mine detectors was based upon the principle of *electrical induction:* any conductor, including all metals, will establish an electromagnetic field when charged. The detectors had both a radio transmitter and a receiving instrument. Transmission of the radio wave induced the flow of current, and thus generation of an electromagnetic field, in a distant metal object. The resultant field could then be detected by the receiver.

STATE-OF-THE-ART DETECTORS

Early mine detectors, the forerunners of the modern metal detectors, were clumsy and heavy, and required bulky battery power sources to generate a sufficient signal. Their sensitivity was minimal and their effectiveness limited to large objects.

After the war, hobbyists who had acquired surplus military detectors could be seen walking beaches and backyards while harnessed into the clumsy contraptions. Most observers smiled and offered condescending nods.

By 1950 a number of firms were manufacturing commercial models considerably less wieldy and more sensitive than the earlier military instruments. But these models still relied upon the same basic components. Arrival of the transistor brought a great reduction in the size and weight of the detectors, and competitive research between commercial manufacturers resulted in the highly sensitive instruments of today.

Detectors now use the most advanced state-of-the-art electronic concepts and components, as well as innovative and functional human engineering design features. Many of these highly sensitive, new instruments have been adapted specifically to the requirements of the modern gold seekers. The degree of sophistication is such that the overall efficiency of use is no longer determined by the inherent capability of the instrument, but rather by the operator's ability to make full use of the features.

SELECTION OF SENSING HEADS

Of particular interest to the gold seeker are the variety of *coils,* or sensing heads, available. Coils range from nearly three feet in diameter—the type recommended for the detection of large, deeply buried objects—to small probes capable of responding to the presence of even the tiniest gold nugget. Some coils are waterproof and designed especially for submersion in streams.

Another useful feature found in better units is *discriminating circuitry*. This enables the operator to determine, without digging, the difference between useless junk such as bottle caps and sought-after objects like coins and nuggets.

Of great value to the gold seeker are the *ground-canceling capabilities* of many detectors, overcoming the problems of operation encountered in highly mineralized ground. Advanced instruments are now able to tune out disturbances caused by the presence of both conducting and nonconducting mineral salts. Gold seekers can effectively work placer deposits and detect nuggets when the troublesome effects of the black sands have been eliminated.

Many placer miners now consider a quality detector part of their standard equipment.

NUGGET SHOOTING FOR COLLECTIBLE NUGGETS

Nugget shooting, the use of a detector in or along streams, has proven productive both in locating gravels well worth panning or sluicing and in the outright detection of larger nuggets. Commercial dredge tailings and sections of bedrock exposed through previous mining work are also good areas for nugget-shooting techniques. Much of the popularity of nugget-shooting is due to the premium value attached to many nuggets over a quarter-ounce in weight. Depending upon visual characteristics of color, weight, and texture, such nuggets may be collectible specimens and worth several times the actual bullion value of the gold contained therein.

OTHER PLACES TO GET GOLD FROM ROCK

While actual hard-rock development is usually a major undertaking, modern detectors make every abandoned hard-rock working a potential source of gold. The early hard-rock miners were limited to eyesight and intuition. Many early mines were highly profitable as long as they were able to track and exploit a vein, but when that vein was no longer in sight, the mines were no longer productive. Many gold-bearing quartz veins were erratic in both size and direction, and the disturbing tendency to pinch out was common. Miners would be left with the decision of whether or not to proceed with the costly hard-rock exploration, hoping desperately to intercept the vein again. Miners sometimes missed the veins they sought by a matter of inches or only a few feet and, faced by nothing but barren rock, abandoned their efforts. Such gold-bearing veins within the rock of a mine wall may be located now with a detector. In recent years success has not been unusual. Some veins are exploited by hand-picking with hammers and picks. Other veins are substantial enough that the property is claimed, then sold to a company for mining development.

Detectors are also used successfully to locate free gold contained in mined ore. In all underground mines a certain part of the ore that had been blasted loose never made it to the ore cars for removal, but was spilled and trampled into the mine floor. There are often heaps of abandoned lower grade ores outside mines. Just because the ores are considered to be lower grade on average assay, it does not mean that outstanding individual specimens containing visible free gold may not be found.

The actual quantity of gold recovered from hard-rock

CHAPTER 11: NEW TECHNOLOGY

mines with detectors usually will not be great, but it is the premium specimen value that makes it worthwhile. Today any ore specimen containing visible free gold is a collector's piece. Its real value has little to do with the actual weight of the gold contained in it.

DREDGING UP FORTUNES

Gold deposits previously untouchable now can be reached and exploited with updated mechanical devices, especially since the development of lightweight, portable suction dredges.

The first effective use of suction dredges in working placer deposits occurred about the turn of the century. Early dredges were composed of a heavy gasoline engine bolted to an even heavier pump. In use the pump created a vacuum, or draw, within a hose or pipe, sucking up underwater river gravels and conveying them to the surface and onto a standard sluice. The gravels that were beyond the reach of early placer miners could then be mined. Dredges were first employed in Central America and California on larger streams and rivers where banks and shelves had been extensively worked. Gravels dredged from the bottoms of the active channels proved equal to the richest bank placers.

Even though the early dredges were not easily portable or particularly reliable, they were used with notable success. In 1927 the International Hydraulic Dredge Company mined the headwaters of the Arkansas River in Colorado's high country. The company was not nearly as imposing as its name, with only four partners and an eight-inch suction dredge. Mounting the dredge atop a clumsy wooden barge, the group was able to work the gravels of the Arkansas's center channel that lay beneath four to six feet of frigid mountain water. The gravels were rich but not spectacularly so. Work was simply a monotonous daily routine of dredging up as much gravel as possible. Each ton of gravel contained only about one-tenth of an ounce (about three grams) of placer gold, but on the best days over four hundred tons would go through the sluice.

Even at the then-undervalued price of gold, the group washed out six hundred dollars in placer gold on their good days. Few men in 1927 earned wages like that. Their gold had come from a region extensively mined from 1860 till it was abandoned in 1865. The gravels had been supposedly mined out.

By 1960 manufacturers were producing complete suction dredge units designed specifically for mining the streams and rivers of California gold rush country. These units were highly portable, compact, and, because of the nature of their operations, required only a lightweight aluminum sluice no longer than three feet. The thorough mixing and washing the gravels received within the extreme turbulence of the hose assured absolute fluidity before they ever reached the sluice. Any gold particles present were immediately trapped in the riffles. The recovery efficiency of the dredge-sluice combinations was the highest ever in the long history of production placer mining. Operators controlled the intake nozzles of the hoses from the surface, working both streams and rivers to depths of as much as five feet. No longer were laborious diversion and damming necessary—if they were possible in the first place—to reach the gold at the bottom of the active channels.

GAUGING TODAY'S GOLD DREDGES

Gold dredges are classified by the diameter of the suction hose. The smallest size is one and one-half inches; such units are usually powered by small, economical one-horsepower gasoline engines. The entire unit—engine, hoses, pump, and sluice—may weigh as little as twenty-five pounds and be easily backpacked to remote streams.

Dredge sizes become larger in one-inch increments with greater weight and power requirements. Six-inch and eight-inch units require several men for operation but are capable of bringing enormous amounts of gravels from stream bottoms to sluices.

CHAPTER 11: NEW TECHNOLOGY

SCUBA GEARED FOR THE SEARCH

When scuba (self-contained underwater breathing apparatus) made its appearance in the early 1950s, some imaginative gold seekers quickly put it to use in conjunction with the dredges. Some of the larger rivers that drained the California gold fields were ideal for this combination. A diver could now visually and manually control the intake end of the suction hose to selectively and completely work gravels anywhere on the river bottoms. Underwater crevices and pockets on and in bedrock, where gold would most likely accumulate in high concentrations, could now be worked completely and efficiently for the first time ever. The recoveries and profits made from the scuba-dredge combinations were in some cases extraordinary, but the operation was not for everyone. In many of the mountain rivers the currents were swift and dangerous enough to require weights of fifty and even one hundred pounds to hold a diver safely and securely on the bottom. A full wet suit was necessary to protect the diver against the numbing effect of the icy waters. Conditions were demanding, and, although the diver did not have to be a competition-grade swimmer, he did have to be nothing less than expert on scuba.

Some of the largest gold nuggets to come out of California's gold fields have been recovered by divers. Many divers increase their effectiveness by taking submersible detectors to the bottom with them. Their recoveries come from the same sources that spawned the great California rush, but this gold was then beyond the reach of the Forty-Niners and those who followed. Although the scuba-dredge combination originated in California, its effectiveness is not at all limited to that area. The technique has been used successfully in all areas of the Rockies.

OTHER SOURCES OF INFORMATION

Detectors, dredges, and scuba, used singularly or in combination, are invaluable to today's gold seekers. But there is still another resource available that is just as important. It is the gold mine of published practical information about every conceivable aspect of acquiring gold from placer or hard-rock sources. Various agencies of the state and federal governments, as well as some foreign governments, provide full information on auriferous regions, the mechanics of staking and recording mineral claims, areas of current production, effective techniques, and even expected costs. Equipment manufacturers publish and distribute freely still more material, covering not only the features of their own products, but a great deal of practical instructional information as well. Libraries are well stocked with how-to books, particularly about various approaches to placer mining and prospecting. Newsstands offer a selection of treasure and gold-oriented magazines.

One of the most noteworthy of the publications is the *California Mining Journal* (*CMJ*) which rightfully bills itself as "the West's leading domestic mining publication." The emphasis in both editorial and advertising content of the *CMJ* is directed almost entirely at the individual or small group mining operations, both placer and hard-rock. More than three-quarters of the magazine is devoted to advertising. Equipment and services include detectors, dredges, gold and gold concentrate purchases, assay services, mechanized separation devices, amalgamation reports, sluices, and hard-rock gear from drills to mills. An interesting classified section lists used equipment, partners wanted, claims wanted, claims for sale, mining legal services, and anything else remotely related to the acquisition of gold from placer and lode sources.

The gold is still there in the placers and lodes, and now the tools and information needed to recover it are available to all. The cost of the new technology may run a bit higher than the traditional pan and sluice box outlay, but the price of gold still assures that the profits will be greater than ever.

12
NEW PROFITS

The nobility and modern mystique of gold are such that many believe the metal should somehow originate as gleaming bars behind the stainless steel repository doors of central banks or, at very least, come from huge mines under rigid corporate control. To some it seems undignified and even preposterous that gold, the ultimate coinage and the grand commodity, is acquired by individuals simply taking it from the ground. Few of these misinformed folk realize the degree to which individual gold mining has been stimulated in recent years, conducted on a very low-keyed basis with none of the publicity and proud displays of recoveries that were such a colorful part of gold rushes. Yet over one million individuals are now actively engaged in the quest for the gold of the Americas. California alone has over thirty-five thousand active mineral claims on record. State officials estimate that every weekend about one hundred thousand persons head for the streams and rivers draining nearby gold fields. A certain percentage of these weekend gold seekers are amateur panners, but most take the quest more seriously, taking full advantage of the extraordinary profit potential.

The economic basis for this tantalizing profit potential has been explained in the introduction to this book. In the decade since 1970 a miner's costs have doubled with inflation while the value of his recoveries—gold—has increased by a factor of about twenty. To the miner this means he is able to equal his profit of 1970 while recovering only one-tenth the amount of gold. Or, if he should recover the same amount of gold, his profit would be ten times greater.

Along with the miner's economic advantage are three other factors which further increase his profit margin: a unique tax-earnings situation, the monetary nature of gold, and the premium value of specimen gold.

The revaluation of gold in the early 1930s (raised from $20.67 per ounce to $35.00) greatly aided the miners, but it was accompanied by another government action which had the opposite effect: the establishment of the federal income tax. The tax was a new twist to the American adventure with gold. No longer could a miner legally keep all he recovered.

MINERS RESENT TAX

Most miners thought it ridiculous that a government could justify a claim to a portion of the gold they unearthed. The nature of their quest dictated self-reliant individualism. If they were successful, it was not because of government help and concern, but was due to their own perseverance and hard work. And now their hard-won income was subject to a government tax. Today, of course, it is subject to many state income taxes as well. But while others smoldered in passive resentment, the gold miners were in a better position.

When the tax law went into effect, the miners founded a tradition of secrecy about their operations. When questioned about recoveries, most miners winked and replied, "In God we trust," meaning, of course, that only God and themselves really knew how much gold came out of the riffles, and neither party was likely to admit it on a tax form.

During the past ten years I have met many gold miners, and I would not wish to imply that a single one of those ladies and gentlemen ever declared even one ounce of gold less than they actually recovered in a given tax year. It is enough to say that lesser individuals might succumb to the temptation present in their particular income situation.

WAYS AROUND THE RULES

Governmental control over and accounting of individual gold production are difficult because gold is more than a valuable mineral and a tradable commodity. In recent years it has also resumed its monetary function. Most placer gold may be converted to paper currency in receiptless transactions, and most miners do just that to meet their living and production costs. The remainder, (which, incidentally, is legally tax free until it is sold), is automatically a solid investment *in gold itself* without the fees and broker expenses incurred in a conventional investment in gold.

It is not surprising that bank accounts and similar traceable investments play a minimal role in the personal finances of most miners. Fine flour gold in concentrate form must be sold to a refiner, with at least some of it becoming part of the official production. Hard-rock ore, of course, must be sold to milling and smelting facilities in traceable transactions, and most goes on record.

Exactly how much secrecy shadows and protects actual gold recoveries is unknown, but the following figures from the state of Alaska indicate the practice is employed to its fullest advantage. The state estimated that during the summer of 1980 between three and five thousand individuals participated in the quest for Alaskan gold. The official production—that gold fully accounted for and subject to taxation—was approximately fifteen thousand ounces. Interestingly enough, several Fairbanks refiners' purchasing agents report the purchase of gold concentrate containing about forty thousand ounces. State officials admit the actual production may have been as high as one hundred thousand ounces.

SPECIMENS UP IN VALUE

The final boost to the overall profit potential is the steadily increasing specimen value of natural gold, both placer nuggets and hard-rock ore. Most natural gold specimens are placer nuggets weighing from one-quarter ounce to many pounds. The smaller nuggets are prized for jewelry use. They bring a small percentage over their bullion weight value as they are by far the most common form of specimen gold. Nuggets from one-half ounce to several pounds may command prices two to three times as high as their bullion value.

Their market value does not depend on weight so much as on color, luster, texture, and shape. A blackened, amorphous nugget might bring only a little more than bullion value, while a smooth, solid, gleaming nugget of distinctive shape, or perhaps one embedded with white quartz fragments from the original lode deposit will bring whatever the market will bear. Some truly beautiful specimens have been sold for nearly five times bullion value.

The gold fields of the Americas have produced some remarkable nuggets. Chile probably holds the weight record, boasting a startling nugget well over two hundred pounds. Alaska, prior to 1910, reported the discovery of at least six giant nuggets, each weighing over a hundred pounds. California gold fields have produced hundreds of nuggets up to twenty-five pounds each.

Giant nuggets are rare, but nuggets of several ounces turn up with surprising frequency. Their discovery and sale usually go quietly and without publicity. Even though a gold seeker might enjoy the attention that would come from the sale of a beautiful nugget, previously mentioned considerations take precedence. The premium value of a quality four- or five-ounce specimen nugget is such that it would take care of a gold seeker's costs for a long time.

GOLD HUNT PRICE TAGS VARY

The actual dollar cost of a gold hunt in the Americas cannot be refined to a flat statement. There are many variables involved. Much depends upon the personal preferences of the gold seeker himself, whether he chooses to stay at the local Hilton dining on filet mignon or to live in a cabin

CHAPTER 12: NEW PROFITS

or tent eating the game he shot that morning. Living and travel expenses make up the largest part of a gold seeker's expenses.

With the single exception of underground development of a small hard-rock mine, the costs of basic equipment remain surprisingly low. For example, a top-of-the-line metal detector with accessories will cost about four hundred dollars. Small dredge-sluice assemblies will start as low as one hundred fifty dollars but increase to several thousand dollars for the heavier, high-volume units. The popular three-inch size, allowing a miner to wash about three cubic yards of gravel per hour, costs about five hundred dollars. Scuba equipment, if desired, will add on another eight hundred dollars. Should a miner decide that the traditional wooden sluice and shovel are best, as it may be for certain conditions, a mere two hundred bucks will buy the tools and materials to put him in business.

The nature of the placer deposit itself dictates the most effective type of equipment. Even after one hundred fifty years of North American placer mining, the basic attraction is yet there. Regardless of what approach or equipment is used, the costs are cheap in comparison to the possible reward. In the 1980s those relative costs are even cheaper than ever before. Many placer miners enjoy a steady, if unspectacular, profit on the common gold grains and dust that make up their "everyday" production. The proceeds pay the way for continuing the search that may lead to a real strike—a layer of rich paydirt, a glory hole, or that one spectacular nugget worth a fortune.

13
IN THE GOLD FIELDS

The Alaskan bush was the scene of my own initiation into individual gold mining. It happened in the 1970s when the price of gold was sharply on the rise, as were the numbers of those venturing into the bush to mine it. With a knowledge of placer mining more academic than practical, my partner, Tom Hanrahan, and I sought both nuggets of Alaskan gold and adventure. In neither respect were we to be disappointed.

ALASKAN PLACER CLAIM

We settled onto a placer claim in a small creek roughly midway between Anchorage and Fairbanks in the remote bush country north of the Alaska Range. The Kantishna mining district was typical of most other gold mining districts in the Alaskan interior: mile upon mile of low rolling hills covered by thick, spongy muskeg carpets and a complex drainage system of creeks and rivers rushing along the base of every ravine and valley.

Gold could be panned from the gravels of nearly every stream in the area.

With the nearest paved road more than ninety miles

away, the district was totally undeveloped and unspoiled. Lone grizzlies and small herds of caribou roamed slowly across the tundra, Dall's sheep climbed the higher ridges, and wolves were often heard, rarely seen. The rivers teemed with arctic grayling; ducks and ptarmigan nested by the thousands. The only access to the region was provided by one frequently impassable dirt road, and the district itself, almost twelve hundred square miles, featured only a few miles of roughly blazed trails. The dominant landmarks of the district were the spectacular twin peaks of Mount McKinley, the highest mountain in North America, forty miles to the south.

KANTISHNA THEN

The history of the remote Kantishna district is based almost entirely on man's quest for gold. Kantishna was the site of one of the many local strikes that spurred Alaskan exploration and settlement in the years immediately following the Klondike rush. In 1905 Joe Quigley, a sourdough prospector, discovered rich placers on a small creek which he named for the day of discovery—Friday. When still more gold was found on most of the other nearby creeks, a small rush ensued. By 1910 the town of Kantishna had been founded, and the entire district, even though it was one hundred fifty long, hard miles from the nearest major settlement at Fairbanks, boasted a population of over two thousand miners. The prosperity of Kantishna was brief for, as in every other gold rush, the best gravels played out quickly. By 1914 most miners had gone elsewhere, following the endless string of rumors that told of strikes in other districts. As always, a handful hung on to continue working the local creeks, slowly and methodically destroying the remains of the town itself. Since the district was located at an elevation of about three thousand feet, well above the low Alaskan timberline, dearly needed firewood was at a premium. Most of the abandoned cabins ended up in the wood stoves of the remaining miners.

KANTISHNA NOW

The townsite of Kantishna is marked by a misleading town dot on Alaskan highway maps. The town itself has disappeared but evidence of the gold rush days remains in every part of the district. Stream banks are littered with the rotted, bleached remains of sluices and flumes along with rusted-out gold pans and shovels, the once sweat-soaked handles long ago gnawed away by rodents.

Few streams still flow through the same channel they occupied when the gold was discovered. Overgrown, dry trenches, winding their way in a braidlike pattern along the present stream channels, mark the diversion ditches dug by the first miners. Both the active channels and the dry ditches are lined with heaps of boulders manually pitched from the gravels before sluicing. Still standing on the hillsides and ridges are the pyramids of rocks that once marked corners of gold claims staked nearly eight decades ago.

Every summer the district is visited by a small group of miners who continue to wash gold from gravels that are far from mined out. While the gold price rise of the 1970s stimulated gold mining all over, not more than fifteen or twenty separate placer operations were active in the entire Kantishna district. Several were run by just one man, most by two partners, and none by more than three miners.

There were only two truly mechanized operations. One group used a small bulldozer to blade gravels into large sluices. Three miles away on another creek, two partners employed a suction dredge to work gravels from the bottom of a rocky chasm. All other operations used only the traditional sluice boxes and shovels with the help of small dams and diversion ditches just as had been done in the gold rush days.

Kantishna's active mining season is unfortunately short, climatically limited to the four months following the spring melt and run-off in late May and ending with the freeze-up in early October. To the miners, time was gold. The long daylight hours of that Alaskan summer saw most groups working twelve- or even sixteen-hour days.

EVERY MAN FOR HIMSELF

Contact between the miners was absolutely minimal. Miles often separated miners from their nearest neighbors. It was probably just as well, for the prevailing attitudes were secrecy, protectiveness, and undisguised distrust. Every miner packed a .44 magnum revolver or a big-bore rifle, ostensibly to serve as protection from the wandering and sometimes troublesome grizzlies, but also because the nearest authorized law and order was about one hundred miles and a full day away. Just as it had been in the gold rush days, everybody looked out for his own ass. Considering the quantities of gold hidden away in and near cabins on the isolated streams, it was a wise precaution. Two years earlier in another nearby mining district, violence that followed attempted theft developed into a full-scale shoot-out. Law and order was finally established by the originally victimized group because they had more guns at the ready. Funeral services were conducted in the bush and mining resumed.

The Kantishna miners had not only themselves to take care of, but a good deal of gold as well. Summer production lent much credence to the belief of geologists and mining authorities that three-quarters of the world's placer gold still remains in the gravels.

Exactly how much gold was washed out during the short season will never be known for, to my knowledge, no Kantishna miner ever disclosed his honest production even to another miner, much less to an official government agency.

In mid-September one miner discreetly displayed what he claimed was his seasonal production, a highly refined concentrate that filled six large Mason jars. Grains and bits of gold were scattered throughout the gray-black concentrate, gathered at the bottom in plainly visible yellow layers. Some of the gold was in half-ounce nuggets and larger. Typical of the placer gold of the district, many of the nuggets contained bits of white quartz from the original lode deposits. The question of how many pounds of gold might be in those Mason jars was answered only by a noncommittal, thin smile.

CHAPTER 13: IN THE GOLD FIELDS

IDEAL CONDITIONS FOR GROUND SLUICING

Many of the district's streams were small and on a fairly steep gradient, conditions ideal for ground sluicing. Plywood dams were constructed a short distance upstream from the section to be worked, below which point the sluice was set firmly in the stream bed. In the narrow, rushing streams, closure of the dam created a four- or five-foot-deep reservoir in ten or fifteen minutes. While the reservoir was filling, the miners picked the larger boulders from the gravels of the now-dry stream bed and pitched them aside.

Release of the dam sent a cascade of many tons of water roaring down the steep gradient of the channel to wash away banks and move heaps of gravel toward the funnel-shaped mouth of the sluice. Repetition of the process allowed the washing of a tremendous volume of gravel with the water doing nearly all of the work.

Another advantage of the ground sluice technique was that bedrock was left cleanly exposed. In the Kantishna district, bedrock was most commonly a chlorite schist that had undergone extensive geological folding. Some of the schist had undergone a ninety-degree shift of the originally horizontal strata and was now aligned vertically. This position made the cracks and crevices of the strata a perfect natural trap for placer gold. One particular section of the bedrock exposed on Friday Creek, the original Kantishna discovery site in 1905, made a sight I will never forget. Hundreds of bits and pieces of gold could be seen gleaming from the deep vertical cracks. Many of the larger pieces weighed about a gram or two, meaning only thirty or fewer would make one Troy ounce.

Although plainly visible once bedrock had been exposed, the gold proved very difficult to extract, the reason why Joe Quigley and his partners had left it there. Today's price of gold certainly warrants mining this deposit, but the narrow, twisting contours of the rocky creek rule out mechanical earth-moving equipment and the depths of the bedrock cracks make dredges largely inefficient. The gold will be mined in time, but for now the claim holder is content to take a break occasionally from his regular ground-sluicing operation and use fine tweezers to fill a small glass vial with a few hundred dollars' worth of placer gold.

HARD TIMES, HIGH TIMES

Kantishna placer gold was notably coarse due to its geological youth and its proximity to the lode sources. The lode deposits were quickly found by the first miners who tore the visible white quartz outcroppings to pieces with picks and dynamite. The ore was then crudely crushed and washed in regular placer sluices to recover the visible gold. Hard-rock development was never attempted in the early years because of the remoteness of the region and the lack of transportation. The inevitable attempt to exploit the lode deposits was made in 1938. A narrow, horizontal tunnel was driven into the side of a ridge to mine the major vein, easily located by following a series of exposed outcroppings. A small mill was constructed to crush the ore and separate the gold. Production from both the mine and the mill began in 1939. Considering the richness of the quartz ore, the operation probably did very well.

Prosperity was short-lived, however, for the gold mine closure order of World War II put a sudden stop to production in 1942. Though the order was lifted at the war's end, the operation was never reactivated. The long, narrow tunnel that led to the gold veins had collapsed and the mill, although still standing, housed nothing but rusted, seized equipment that would never operate again.

The story of the mine and mill did not end there. In the early 1960s a gold seeker packed into the district, bringing with him in several arduous trips about a hundred pounds of mercury. For three weeks he worked in and near the old mill, dismantling equipment and cleaning dirt and grime from the crushers, separation shaker tables, ore bins, chutes, and conveyor belts. From the floor of the mill he collected the spill accumulated during the brief years of operation. By the time he was finished, his mercury weighed over one hundred fifty pounds and looked like a dirty sponge.

CHAPTER 13: IN THE GOLD FIELDS

Through amalgamation he had recovered about fifty pounds of gold for which he received sixteen thousand dollars, an impressive reward for a few weeks of work in 1963. While that was certainly worth his while, it happened at a time when gold was notoriously undervalued and the true age of the individual gold seeker was still years away. Today the gold-mercury amalgamation would be worth at least a quarter-million dollars.

QUALIFIED SUCCESS

During the short Alaskan mining season, I estimate the fifteen or so operations in the Kantishna district took about one hundred pounds of gold from the sluices. Every miner benefited by the climbing price of gold. My partner and I shared in the district production, sluicing our gold from the gravels of a small creek never wider than six feet. Our equipment and materials were limited to picks, shovels, a few sheets of plywood, and basic woodworking tools. Expenses were minimal, recoveries good—the quest was successful.

Not to be forgotten is the fact that the gold did not lie on top of the gravels, simply waiting to be picked up. Had that been the case, the miners in 1905 would have saved us the trouble. There was a great deal of physical work under conditions not always pleasant. Months of isolation were involved, for to make it pay, miners must go in to the remote districts at the first thaw and come out at the freeze-up. May, June, and July can be counted on for cold rain and clouds of mosquitoes. The water in the creeks, where a miner will spend much of his time, will rise above forty degrees Fahrenheit only in late August. Disappointment must be accepted readily, for often after the work has been done, the hoped-for layer of paydirt has eluded you.

But the gold is still waiting in the gravels—of *that* there is no question. With perseverance, much satisfaction may be realized with the final washing of the sluice concentrates. You will know the work is worth it when the last of the black sand concentrates is poured off and a swirl of the remaining concentrate reveals a brilliant yellow streak of gleaming, coarse placer gold stretched halfway around the pan.

To a man, the miners accepted the hard work and months of isolation, content in the knowledge that the gold they washed out, at its greatly increased value, would pay for the rest of the year. For most, summer mining had become the primary livelihood. Some still shook their heads in disbelief at their rising good fortune tied to the price of gold. Their costs, over ten years, had increased only moderately, while their incomes had increased dramatically, enough to put many in what some might refer to as a high tax bracket. But not many miners worried about tax brackets.

One veteran Kantishna miner, who had doggedly worked the district streams during the lean years when gold was undervalued, but who also had stuck it out to enjoy the profits he makes today, said it all:

"Well, hell, since a man's gotta dig for somethin', it may as well be gold."

And with a backward glance at his claims as he was packing out in late September, his final comment was, "Just like goin' to the bank."

Perhaps, even a bit better.

14
NEW GOLD RUSHES

The greatly increased levels of activity in the gold fields of Alaska and other areas of North America cannot be considered true *gold rushes* in the historic sense of the term. Yet the last of the gold rushes of the Americas did not fade into the past with the Klondike; it is just that gold seekers must now journey a bit farther to join them. A number of regions in South America that once yielded enormous quantities of gold to the early Spanish and Portuguese are again the sites of frenzied mining activity. These qualify as full gold rushes in every way.

GOLD DEEP IN BRAZILIAN JUNGLES

Small-scale gold mining in the vast jungles of Brazil has been carried out continuously ever since the Portuguese made and exploited their famed strikes in the colonial years. During the 1960s modern prospectors drifted through jungles that remained unsettled and, in some areas, even unexplored and unmapped. A few local, rich strikes were made on small placer deposits which, located anywhere else but in the depths of the Brazilian jungles, would have drawn much attention. The rapidly increasing price of gold during the 1970s lured more prospectors into the remote jungles. Finally, a tremendous gold strike was made in January 1980 at Serra Pelada in the southern part of Brazil's Para State.

The Serra Pelada strike was not something that merely encouraged further exploration. This was the real thing: gold in quantities that rivaled the two-hundred-fifty-year-old Portuguese discoveries. Reaction to the excited reports of a gold strike had changed very little over the centuries. The local mahogany industry ground to a halt as workers left the lumber mills. Ranchers watched helplessly as their *peoes*, hired hands, collected their pay and walked off the job. Within one week thousands of hopeful gold seekers were headed through the jungles for Serra Pelada.

Few of the early arrivals were disappointed. They were able to stake choice claims under a system similar to the mineral laws of the United States. During January and February 1980, with gold riding at over five hundred dollars per ounce, the ounces were washed from the gravels by the tens of thousands.

Some miners who arrived at the diggings with nothing but the shirts on their backs found themselves the proud, and suddenly wealthy, owners of ten or twenty pounds of gold each. A number of fiteen-pound nuggets turned up which, sold at bullion value only in the excitement of the rush, were worth over ninety thousand dollars each. The ex-lumber cutters and cowhands were awed, of course, by the richness of the gravels. Everyone worked feverishly through the day. Many worked through the night by the dim light of kerosene lamps, and for good reason. For most of them, a single one-pound nugget represented five full years of laborer's wages.

The details of the Brazilian discovery are already local legend. Like the stories of discoveries that started the United States rushes, they, too, may be a bit over-romanticized.

The most popular story tells of a great chestnut tree felled in a storm. When the tree had fallen, gold nuggets glittered in the exposed root system. Sounds like the South American version of the Georgia deer that kicked up the gold nugget.

While that story might be questioned, the richness of the deposits could not. At the end of the first ten months, twenty thousand miners sweated away in what had become a vast open pit. Their total production had reached two hundred thousand ounces (about eight tons of gold) with a value of about $100 million. The magnitude of the numbers may automatically recall visions of the gold rushes of long-ago history, but they apply to the *first ten months of 1980.*

LIFE AND DEATH IN SERRA PELADA

Law and order was nearly nonexistent during the early chaotic months at Serra Pelada. What little authority there was belonged mostly to the gun-toting *congacieros,* the gangs of jungle outlaws that used violence to extort gold tribute from the miners. The Brazilian government stepped in with a form of loose martial law similar to that which the Canadian government used to keep the lid on the Klondike in 1898.

Serra Pelada has since developed into a ramshackle boom town of open-walled huts lining muddy streets. The settlement is surrounded by devastated jungle where trees have been cut and sawed into timber to meet the miners' lumber demands.

The late arrivals unable to stake their own claims are employed as day laborers. For a wage of twenty to thirty dollars per day, they haul gravel from the bottom of the mile-wide open pit up a maze of as many as six rickety wooden ladders to the sluices where it is washed.

Since July 1980 a Brazilian government order requires all gold to be sold to the Federal Economic Bank. The bank agency established at Serra Pelada is a low, frame building where seven khaki-dressed purchasing clerks, protected by military guards, purchase any and all gold for 75 percent of the current London price quote. Pay windows are classified according to the size of the sale, with wooden signs lettered in Portuguese indicating "Less than 100 Grams," "200 to 1,000 Grams," and "Over 1,000 Grams."

A number of rags-to-riches stories have come out of the Serra Pelada rush, but the most remembered is that of

Geraldo Joto, a thirty-four-year-old ex-grower who was having little luck with his small, break-even farm. When he heard of the strike, Joto sold his farm to the first taker, headed for Serra Pelada, and invested the meager proceeds of his farm into his own claim. His claim was one of the richest of the entire rush. The gold Joto washed from his gravels was more than enough to purchase a twenty-five-hundred-acre ranch stocked with one thousand head of prime beef cattle. In a matter of a few months, Joto had mined four hundred kilos (eight hundred eighty pounds) of coarse gold nuggets.

The 1980s hold every promise of being a golden decade for Brazil. Strikes in the general Serra Pelada region have occurred since the original discovery. The richest of the satellite rushes was founded on another imaginative strike, this one when a large gold nugget was seen wedged in the treads of a tractor. One might question this account also, but the six thousand gold seekers participating in the following rush could care less. They are too busy washing out one hundred pounds of gold per week from the "Mine of the Tractor."

This succession of new strikes is expected to continue for several years, as there are several hundred thousand square miles of the Brazilian interior *never thoroughly prospected*. This optimistic prediction is made not only by the thousands of gold seekers who dream of duplicating the success of Geraldo Joto, but also by professional geologists who base their opinion on encouraging general regional surveys.

So great is Brazil's gold production now that it will soon surpass that of the United States and Canada and rise to third in overall world production. While claims are limited to Brazilian citizens, the rushes have become international in scope, with many foreigners grubstaking citizens for mining and prospecting operations.

WHERE IT'S AT IN COLOMBIA

Sharp increases in the level of individual gold mining have also been reported in the Republic of Colombia, with local rushes taking place. Unlike Brazil a number of Colombian regions have become big gold producers not because of new strikes, but simply because the current price of gold has lured prospectors and miners back to the same deposits left by the early Spaniards. Among the top producing regions now are the entire western slope of the Andes along the Pacific Coast, the long Cuaca River Valley, Antioquia Province, and Choco Province near the Panamanian border.

The largest single rush was to the La Sierra gold fields in Antioquia where eight thousand gold seekers are now working both placer and lode sources. The placers are worked by every method ranging from bateas to sluices to modern suction dredges. Hard-rock production comes mainly from a maze of crude and extremely dangerous tunnels and shafts. The gold must be sold to the Banco de la Republica, the Colombian government bank, which has recently had to substantially increase its staff of purchasing agents to handle the new production. La Sierra's production in 1980 alone amounted to some twenty-four hundred pounds of gold worth about $17 million. The importance of the *individual* gold seeker in the Colombian rushes is clearly reflected in recently released official figures. Ten years ago the Colombian central bank purchased only 20 percent of its gold from individual prospectors and miners. In 1980 the greatly increased Colombian total production reached nine tons or about $100 million. The central bank purchased *80 percent* of that fortune from individual prospectors and miners.

RISKS OF THE QUEST

Internal political violence and traditional banditry common to Colombia make gold mining a risky adventure. Every year a growing number of miners are killed and injured in robberies, claim jumping, and raids by guerrillas after gold to finance their own activities. In one notorious case, gold discovered on Indian lands near the Panamanian border cost the lives of both Indians and police.

While the risks of the quest for gold in Colombia are serious, they are not enough to deter the gold seekers, many of whom are making fortunes. The future of mining in

Colombia is gold plated also, for the central bank expects continued rises in production over the coming years.

BLACK MARKET GUNS IN GUYANA

The People's Cooperative Republic of Guyana on South America's northeast shoulder has its own productive gold fields, but the miners there work under a different set of "rules" because of a unique political-geographical situation.

Since the late 1960s Guyana has been governed as a socialist nation. The policies instituted have since succeeded in destroying the economy, eroding the confidence of the people in their own government, and making the Guyanese dollar worthless on the foreign exchanges. Almost the entire population of Guyana is concentrated along a narrow coastal section. The remainder of the country is roadless jungle stretching nearly four hundred miles south to the Brazilian border. It is generally in this remote, isolated region that ancient gold placers have been found beneath muddy jungle rivers.

A small society of rough-and-ready gold miners known locally as *tommyknockers* have worked the rivers for decades using suction dredges, the only really practical method. In recent years their numbers have grown considerably along with the price of gold. Most tommyknockers are Guyanese nationals, although few even acknowledge the existence of the Georgetown government far to the north of the gold fields. Government control of the gold fields is limited to a handful of fortlike outposts from which the police and military manage infrequent, brief patrols into the jungles.

Guyanese law requires that the tommyknockers sell their entire gold production to the government in return for Guyanese paper currency. The miners, who work hard and put up with months of deprivation and isolation in the jungles, regard this regulation with amused contempt. The actual amount of gold production is very uncertain, for the government sees only a fraction of it. Most is spirited across the unpatrolled borders to Brazil or west to Venezuela where it is eagerly purchased for a much more convertible currency,

CHAPTER 14: NEW GOLD RUSHES

or disposed of on Guyana's rampant black market.

Getting to the Guyanese gold fields is a job in itself. The first step is a journey by decrepit river boat some forty miles up the mile-wide Essequibo River.

Gold seekers disembark at the frontier town of Bartica where the main industry is providing the equipment and supplies for the gold miners who work much farther into the jungles. The black market thrives here, selling illegal guns to the tommyknockers and buying their gold with foreign currency or, at greatly inflated values, with the local currency.

Although actual production may never be known, it is said that three tommyknockers working a six-inch dredge wash out rarely less than an ounce or two of gold every day. Knowledgeable black market entrepreneurs estimate the total Guyanese production to be as much as two thousand pounds of gold worth about $10 million.

The individual quest for gold from the rock did not die with the Klondike. All through the Rockies, from Alaska to South America, gold production in recent years has been greatly stimulated, reaching all-time highs in many areas. And more of that gold is being mined by individuals than ever before; most are realizing excellent profits, a few are taking out fortunes.

The lodes and placers had not been mined out. Only a realistic adjustment in the price of gold was necessary to lure men back to the gold fields.

The attitude of secrecy developed among treasure salvors holds true today more than ever. A standing rule of the trade is "if you found it, tell 'em you didn't, and if you didn't, tell 'em you did."

III
FROM
THE
SEA

15
SEA-GOING GOLD

Those first bits of New World gold that Columbus acquired were both beautiful and promising but, at the same time, rather worthless in the European sense. For its full and expected value to be realized, the gold had to be transported across the Atlantic to Spain and other economically depressed European economies. In the first century after Columbus's discoveries, any trans-Atlantic voyage was a matter of considerable risk, but one that had to be taken if the new-found gold was to achieve full power and monetary value.

SPANISH TREASURE FLEETS

Spain's first shipments of gold were small, consigned to whatever reasonably seaworthy vessel was returning eastward. That method sufficed until Cortes and Pizarro found gold by the ton. Even more gold came from the rich mines. Spain then found itself confronted annually with a mammoth logistical problem.

The solution came in the form of a regular shipping system. Annual treasure fleets of heavily armed galleons collected the gold and silver of the New World and transported it across the Atlantic. Virtually every bit of gold acquired during the great conquests or mined during the centuries of exploitation, sooner or later, became subject to the hazards of sea transportation.

After the demise of Spain's colonial empire, much of the mined gold remained in the Americas, possibly to be shipped later as part of regular trans-Atlantic trade transactions in a safer maritime era. But from 1500 to 1820 the Spanish attempted to ship as much as fifteen hundred tons of gold in an estimated seventeen hundred easterly Atlantic crossings, an impressive register of voyages to which the Portuguese contributed a few thousand more. Added to all of those were a great number of French, British, and Dutch voyages. While those European powers mined and shipped nothing on the scale of Spain and Portugal, many of their ships still carried quantities of gold acquired either by trade or by force. By 1820 the five European powers with major New World interests had made over one hundred fifty thousand voyages to and from the Americas.

A LOT OF LOST GOLD

Compared with modern standards, colonial shipping losses were nothing short of staggering. Each and every voyage pitted both crew and ship against a combination of dangers. Gold shippers had to take into account the decrepit condition of the wooden hulls, the uncertainties of early shipbuilding and navigation techniques, the notoriously inaccurate charts, the vagaries of weather, the omnipresent threat of enemy or pirate attack, and the treacherous nature of the waters through which many of the voyages were made. The latter by far claimed the most ships as the route went through the shallow, reef-lined waters of the Caribbean region.

At least 5 percent of all colonial shipping was lost, historians estimate. Possibly fifty tons of Spanish gold never made it to Europe. Most of it still rests with the long line of shipwrecks throughout the Caribbean region.

CHAPTER 15: SEA-GOING GOLD

To anyone but a maritime historian, the ships that carried the gold of the Americas may seem incredibly small and fragile. The *Santa Maria* was only about seventy feet in length, typical of the era not only in size but in design. Columbus's flagship featured the basic configuration that the Spanish would favor for centuries, one that had an unusually deep draft relative to the overall length.

The *Santa Maria* was lost when it brushed its worm-weakened hull against a coral reef, demonstrating a problem that would plague wooden hull mariners for centuries. The Caribbean's toredo worms, with their insatiable appetite for wood, could reduce a new wooden hull to little more than a mass of softened timber too weak even to hold spikes in a matter of months. The coatings of pitch first applied as a deterrent proved a temporary measure at best. Copper and lead sheathing were later used for protection, but wooden hulls still proved vulnerable to the attack of the worms. A number of ships constructed in Caribbean harbors actually sank before embarking upon their maiden voyages when pumps could not handle the leakage through worm-perforated planking.

Sails and hemp line also rotted in the warm Caribbean climate. Food and perishable stores quickly spoiled in the high temperatures and humidity, as did the holding casks of precious water. Crews sometimes became so debilitated from the effects of spoiled food, bad water, and vitamin deficiencies they were barely able to man the ships in calm seas, much less in emergency situations.

By the year 1600 some of the larger ships exceeded one hundred feet in length while retaining their bulbous hull configuration. Cargo capacity was increased greatly, but so was the draft, which often reached twenty and sometimes nearly thirty feet, more than enough to assure striking even the deepest reefs. While cargo capacity was a major consideration of Spanish mariners, maneuverability was not. The galleons and merchantmen of Spain were very poor in tacking and steering ability. When shipbuilders created improvements in design, it was the British, French, and Dutch who would benefit, not the Spanish and Portuguese.

Inadequate navigation, or rather the lack of accurate techniques and charts, presented another serious hazard to colonial shippers. Sun sightings and the magnetic compass, understood in the time of Columbus, remained the primary tools of navigators for centuries. The magnetic compass allowed a navigator to determine direction relative to a variable and still-mysterious magnetic north, and sun sightings provided a reliable measurement of latitude. Longitude, the determination of relative east-west position, was impossible to establish at sea until the introduction around 1730 of accurate shipboard timepieces. Charts drawn prior to that time show significant longitudinal error concurrent with remarkable latitudinal accuracy. The problems of determining longitude, together with errors associated with dead reckoning navigation, brought many ships to grief upon the reefs.

By the mid-1700s the use of the chronometer and the sextant made for safer sailing, but all shipping remained at the mercy of the weather. The terrible hurricanes that swept the Caribbean in late summer and fall were totally unpredictable. Ships caught in the great storms might survive if they could run before the wind on open sea. But the shipping routes of the Caribbean did not allow such open expanse of sea as most routes were bordered by a mainland coast on one side and coral reefs on the other.

Considering all the hazards early mariners faced on every voyage, it is remarkable that more ships—and more of the gold of the Americas—were not lost.

GOLD ROUTES FROM THE NEW WORLD

The early trading routes were dictated by the trade winds. The voyage from Europe to the New World was quickest if ships first sailed south along the northwest shoulder of Africa, then picked up the easterly trade winds that would bring them to the New World near the southern end of the West Indies. Return voyages to Europe were facilitated by leaving the New World near the Bahamas and following a route across the North Atlantic.

CHAPTER 15: SEA-GOING GOLD

Routes within the Caribbean were established on a compromise between the prevailing winds and the locations of major ports. In South America the most important treasure port was the fortified city of Cartagena, Colombia, where all the production of the continent was accumulated.

The treasures of Peru and Chile were shipped northward along the Pacific coast of South America to Panama where they were offloaded and transported across the isthmus by mule. At Panama's Caribbean port of Portobelo, they were loaded aboard ships for the short voyage to Cartagena. There, beneath the massive stone fortress walls and protection of the harbor guns, Peruvian treasures were consolidated with the gold and emeralds of Colombia. All was loaded aboard the waiting fleet for the voyage north to Havana, Cuba. This route took the treasure galleons through one of the most dangerous regions of the entire Caribbean.

About four days north of Cartagena, the treasure ships passed perilously close to a series of isolated coral reefs marked only by low-lying sand islands about one hundred fifty miles east of the mainland coast of Nicaragua. By the mid-1500s the Spanish had given names to these deadly reefs. Serrana Bank had been named for an unfortunate Spanish sailor who had been cast ashore there to live a Robinson Crusoe type of existence for eight long years. *Serranilla,* or "Little Serrana," was an adjacent reef. *Roncador,* the "snorer," was named for the ominous thunder of the breakers that crashed continuously upon the coral. *Quito Sueno,* literally "stop dreaming," was in its own name a warning to the Spanish lookouts. Sun sightings told Spanish mariners when they were in the latitude of the reefs, but estimation of longitudinal position was open to the errors of dead reckoning.

Very aware of their navigational and maneuvering deficiencies, early mariners routinely carried ten or twelve large anchors rigged for instant deployment, the last hope of holding a ship off the coral when it was no longer possible to steer or sail clear. Numerous wreck sites have been located on these reefs today by following the trail of anchors, each of which failed to hold the bottom or was lost when lines separated.

A second treasure fleet also sailed for Havana, this one departing the harbor of Vera Cruz, Mexico, with the gold and treasures of Mexico and sailing a circuitous route northward into the Gulf of Mexico. Beneath the protective walls of Havana's great Morro Castle fortress, the fleets repaired their vessels and took on supplies needed for the final leg of the journey that would deliver the gold of the Americas to Spain. When the combined treasure fleet departed Havana, the last danger lay in safely clearing the remaining land mass of the Americas. Once the ships were in the open Atlantic, arrival in Spain was almost a certainty barring an unfortunate encounter with a severe storm.

Prior to 1560 the fleets followed a particularly dangerous route that took them eastward along the northern coast of Cuba in what is now known as the Old Bahama Channel. At its narrowest the channel offers only twelve miles of deep water as it threads its way between the reefs of the Bahama Bank and Cuba. Even after successful navigation through the Old Bahama Channel, the fleets had to contend with two equally dangerous passages leading through the Bahamian chain itself before reaching the safety of the open Atlantic.

After 1560 a new route was established which made use of the strong current of the Gulf Stream. From Havana the homebound treasure fleets turned northward with the Gulf Stream, sailing between the mainland of Florida on the west and the reefs of the Bahamas on the east. The open Atlantic was reached when the most northerly of the Bahama Islands was cleared. The reefs of Bermuda, far to the northeast and a pinpoint in the vast ocean, remained the last obstacle. These reefs, too, claimed their share of ships.

FIVE THOUSAND COLONIAL WRECKS

The routes used by Spain's treasure fleets in the waters of the New World totaled over seven thousand miles. Today, though the great galleons no longer sail them, they still have one thing in common: they are littered with the bones of lost ships. After the end of the colonial era, the volume of ship-

CHAPTER 15: SEA-GOING GOLD

ping throughout the Caribbean region increased with the expansion of general maritime trade. Losses continued with regularity but at a reduced rate as sailing slowly grew safer. During the 1800s most warships and merchantmen commonly carried quantities of gold coins for payroll and trade purposes. Accordingly, any old wreck site in the Americas today is a potential source of gold.

Many wrecks, particularly those in the Caribbean region, because of shallow water and generally favorable conditions, are well within reach of the individual gold-seeking salvor today. It is estimated that Florida and the Bahamas alone may have as many as three thousand old wreck sites. Throughout the entire Caribbean region, that number may be as high as five thousand.

NUGGETS, COINS, AND MONEY CHAINS

The gold of the Americas was shipped in every form imaginable. In early colonial years, granules and nuggets directly from the placers were shipped from Hispaniola. A few decades later the Spanish treasure ships were carrying examples of the exquisite golden artwork of the Central and South American Indians. Most of the gold the Spanish shipped from the Americas was in the form of crudely cast bars and disks of various sizes, many bearing impressions indicating purity, weight, proof of tax payment, mint of origin, and sometimes dates and consignment numbers.

By the mid-1600s the Spanish colonial mints were striking gold coins in various demoninations such as the one-, two-, four-, and eight-escudo pieces. The eight-escudo coins were also known as *doubloons.* They have since become a romantic symbol of the era of the Spanish Main. Chests of gold coins were carried on the eastbound treasure ships along with large amounts of personal jewelry such as rings and religious objects. Since jewelry, if worn upon debarkation in Spain, was excluded from the crown tax, money chains became popular. Often of absurd length and weight, the chains had hundreds of gold links of uniform weight. The links could be used conveniently for money while the wearers cleverly avoided the tax. In the later part of the colonial era, gold coins of other powers, particularly England and Portugal, joined the Spanish coins as a universally accepted medium of exchange for trade and were common items aboard most ships.

SMUGGLING AND COUNTERFEITING

While Spain did its best to regulate completely the shipment of gold, the greed and imagination of the Spaniards often made this impossible. All Spanish treasure shipments were recorded in utmost detail to insure accurate accounting of every piece when it reached Spain. Still smuggling flourished as the gentlemen and soldiers of Spain made every effort to deflect the tax bite from their personal golden wealth. So substantial was the smuggling that some wrecksites have yielded nearly double the amount of gold listed on the official manifests. Illegal (untaxed) gold was secreted away aboard the treasure ships in every possible place. One of the more innovative hiding places was down the barrels of the deck cannon, counting upon the mighty assumption that the particular gun would not have to be fired on the return voyage to Spain.

The Spanish gold was also subject to counterfeiting as the colonial assayers and mint workers found themselves in a position to amass personal fortunes very quickly at the expense of the crown. By the mid-1700s counterfeiting of gold coins had reached grand proportions because of the easy availability of the ultimate counterfeiting medium—platinum. As much of the "worthless" platinum as possible was alloyed with the gold without substantially changing the yellow color. The replaced gold, of course, went into the pockets of the Spanish mint workers.

All the gold ever lost in shipwrecks in the Americas—whether English, Spanish, or Portuguese, legal, smuggled, or counterfeit—has two common characteristics. First, when it is recovered it is every bit as clean and brilliant as the day it was lost. Second, no matter what the form, whether coins, bars, or jewelry, its value is many times that of the bullion weight of gold, thanks to growing numismatic and artifact values. Upon these two qualities is founded the lure of treasure salvage.

IN SEARCH OF GOLD

More gold than ever—with a higher value than ever—is being recovered right now by individual gold seekers. The profit potential, that difference between the real purchasing power of gold and the costs of the quest, has never been greater. This means that a small amount of gold, such as this Spanish doubloon salvaged off an ancient wreck site (*above and on cover*), may recoup the expenses of the search. The doubloon is actually about the size of a silver dollar. The *M* and *J* markings to the left of the shield indicate the coin's minting in Mexico and the assayer's identity, Jose Eustaquio de Leon. To the right of the shield, the *VIII* tells the coin's original value in escudos. Its present-day value in dollars is another story. The numismatic value of historic coins and other gold artifacts means they command prices over those prescribed by their gold content.

From the Rock

The basic tool for placer mining is the sluice box (*above*), nothing more than a modified and controlled stream bed in miniature. *Right top to bottom:* More tools of the trade include those used in the separation of the gold.

FROM THE SEA

A disk of Spanish gold (*above*), gleaming as brightly in 1979 as on the day it was poured, flashes in the sun off the coast of Florida. The disk was recovered from the wreck site of a 1622 treasure galleon, the *Atocha*, found about thirty-five miles west of Key West, Florida. Two separate salvage companies worked the wreck; Mel Fisher's Treasure Salvors, Inc. and Dennis Standefer's Seaborne Ventures, Inc. Pictured aboard Standefer's salvage ship, the *R/V James Bay*, is treasure diver John Doering, who helped bring up over $1.1 million in gold. *Opposite top to bottom:* Included in the recoveries were many stamped bars, a ten-foot-long money chain worth over a quarter-million dollars, and a ten-ounce gold bar and a delicate chain.

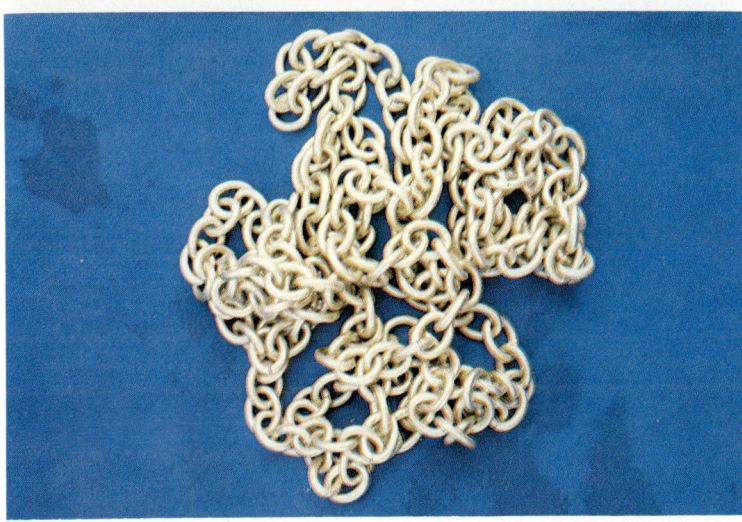

FROM THE GRAVES

The 500-year-old tradition of the *huaqueros*—the once-respected trade of professional graverobbing—endures in Central American jungles. *Above*: Huaqueros and a burro loaded with digging tools go into the jungle. Successful excavations of pre-Columbian burials yield such gold artifacts as a breastplate (*below*), frogs (*opposite*), and a stylized human figure (*next page*).

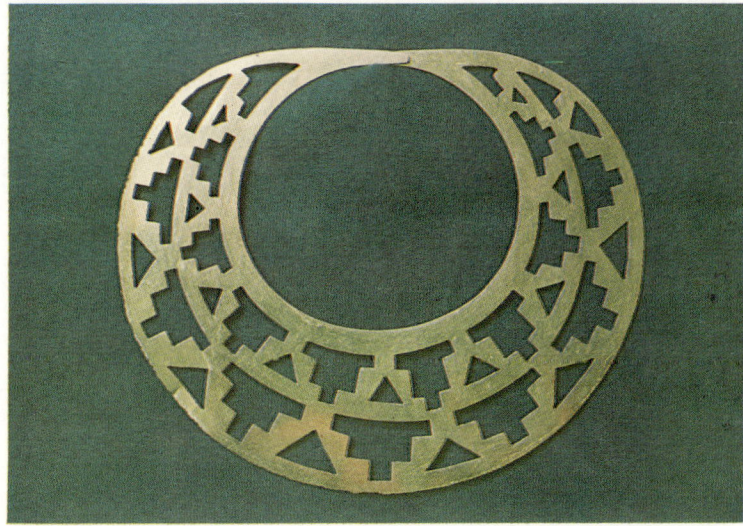

16
EARLY SALVORS

Today's shipwreck salvors of the Americas trace their colorful heritage right back to the crew of the *Santa Maria*, wrecked while testing the mettle of a Haitian reef. While the ship was firmly grounded, everything of value or possible use was stripped and hauled ashore before the work of the tides and waves reduced the hull to nothing. The wreck of the *Santa Maria* also provided the opportunity for the first treasure salvage in the Americas, for Columbus's crew succeeded in recovering the modest amount of Indian gold aboard. Attempts at treasure salvage soon became painfully routine, especially after Cortes's conquest of Mexico. The beginning of the annual treasure fleet departures brought increasing ship losses, and the Spanish found themselves faced with the necessity of recovering a growing amount of sunken gold. As the number of sailings increased, the wrecks became even more common, so did the departure of the salvage expeditions from their home port of Havana.

It was imperative that the salvage efforts begin at the earliest possible time. The Spanish had to recover their lost gold before the passage of another storm which might completely break up and scatter the remnants of the hull, burying the treasure forever beneath many feet of drifting sand on the sea bottom.

The second concern to the Spanish was that each of their wrecks was fair game to the competition, and the English, French, and Dutch, together with a growing legion of pirates and freebooters, would like nothing better than to get first crack at a rich Spanish treasure wreck. The Spanish often lost out to their competition.

The winning attributes of speed, secrecy, and courage would be passed along to succeeding generations of salvors.

CHALLENGES OF THE WRECKSITES

Most wrecks are classified in two categories according to their general condition, *intact* or *scattered*. A sinking in calm seas, perhaps after holing on a reef then drifting away to sink in relatively deep water, would usually result in a basically intact wreck site; that is, the hull remained in one piece with all the cargo contained within. Scattered wrecks were those of ships that struck reefs in heavy seas and high storm winds. The hulls would often break up as they were dragged and pounded across the coral in the heaving waves, resulting in a *spill* or trail of cargo, wreckage, and treasure between the initial point of impact on the reef and the final sinking point of the hull. Virtually all reef wrecks, in time, were scattered to some degree because of the destructive effect of storm tides and pounding surf.

Immediately upon sinking, the ships and everything aboard underwent various forms of biological, physical, and electrochemical deterioration. Every wreck eventually was reduced to an amorphous, anonymous mass on or within the sea floor. As tides and waves broke up the hull, the more fragile objects (such as those of ceramic material or glass) usually were fragmented at least partially. Any wood not completely covered by sand after a sinking was devoured down to the last splinter by toredo worms in as little as ten years. Electrochemical deterioration affected all metal objects from the massive iron cannon and anchors to the more delicate objects of brass, bronze, and silver.

Only the gold, thanks to its extraordinary inertness

CHAPTER 16: EARLY SALVORS

and durability, survived intact through centuries of saltwater submersion.

EARLY UNDERWATER SALVAGE

Considering their primitive tools and techniques, the early salvors managed some impressive feats. The greatest single advantage available was the free diving ability of the Indians, many of whom could work for several minutes at depths down to sixty feet. Simply reaching a wreck, however, in no way assured that recoveries would in fact be made; clearing heavy timbers and attempting to penetrate the hulls were brutal and dangerous operations. Early underwater salvage, especially if any appreciable depth of water had to be contended with, was invariably a long, slow process.

The Spanish worked continuously on some of their own treasure wrecks for as long as ten years before finally abandoning them to the sea and the other salvors waiting impatiently for the scraps. Early salvage was costly, not just in time and money, but also in lives of the Indians forced to perform the diving. As if enough Indians were not killed when the gold was acquired, now others died so that the Spanish might recover their losses.

Only a very few of the early shipwrecks were ever effectively salvaged even if their locations were known. The locations of many others were lost with the ships that struck remote reefs, going down with every person aboard. Others holed on reefs only to drift away into deeper waters to sink, their graves never marked. Still others were shoaled over quickly with no visible trace remaining, but possibly to be uncovered by storms centuries later.

Throughout the Caribbean region today there are many hundreds of wrecks that have never been touched by man since they were lost.

SHALLOW-WATER WRACKERS

Through the 1800s marine salvage still relied on the same primitive techniques used for centuries. Crude diving bells, simply inverted housings capable of maintaining their trapped air supply, brought limited success in certain applications. Unreliable manual pumps were employed to provide air to divers, but such experiments ended in disaster more times than in success. Even though significant technological advances would not be realized until the twentieth century, shipwreck salvages developed into an industry throughout the Caribbean, the Florida Keys, and especially the long chain of Bahamian islands.

The salvors became known as *wrackers* and made their livelihood out of thoroughly stripping everything of value from the hundreds of shallow-water wrecks that littered the reefs. What their simple tools were unable to accomplish was often managed with ingenuity, determination, and courage. Some groups became notably successful in what was a highly competitive profession. The wrackers are best remembered not for their operational proficiency, but for their overzealous and highly questionable methods to insure their own occupational security. Much of their work depended upon the steady occurrence of shipwrecks. During those periods when the reefs claimed no ships, their work and earnings fell off. To stimulate business, the wrackers were not above placing false lights on points and beaches for the express purpose of luring ships onto the reefs and shoals.

Wracking as a full-time profession continued into the 1900s with the salvors working in tight groups, upholding secrecy, and, if necessary, making their own laws to fit the situation at hand. Possession of a new wreck fell to those who physically claimed it first, and maintaining possession until the job was done was often a matter of which group of salvors had the most guns and willingness to use them.

The wrackers performed some remarkable feats of salvage, often dismantling entire ships. But they recovered very little of the lost colonial gold. That was destined to remain in the sea until the arrival of modern times and technology that would allow salvors to work effectively beneath the sea.

17
SALVAGE TECHNOLOGY

Contemporary treasure salvors, like today's individual gold miners, are the most productive and successful in history. The first technological advancements to benefit salvors were the reliable compressed air systems that came into use about 1900. Used in combination with the new hose-fed hard-hat diving helmets, they provided divers with the capability to perform long hours of work at depth in relative safety. Hard-hat gear ushered in the modern era of marine salvage. But because it was heavy, required extensive support systems on the surface, and cost a great deal of money, the gear did little to aid the individual gold seeker, whose time would not come until the arrival of scuba equipment.

WARTIME TESTING OF BREATHING SYSTEMS

Experimentation with a portable compressed air breathing system first happened in World War II. By the early 1950s safe and relatively inexpensive scuba equipment was commercially available. Its acceptance and use among salvors coincided precisely with the beginning of a long, continuous string of treasure salvage triumphs. Salvors were capable of working not only the shallow-reef wrecks, but even those at depths exceeding one hundred feet. The sunken gold of the Americas that had eluded salvors for centuries was suddenly within reach.

AIRLIFT EXCAVATION OF SEA-BOTTOM SEDIMENT

Yet the ability to reach a wrecksite was not enough. Nearly all ancient wreck sites were covered with deep layers of sand, mud, silt, and other sea-bottom sediments. If anything at all was to be salvaged, the wreck site had to be excavated first. Industrial dredges had long been available for such work, but to benefit the individual or the small salvage group excavation equipment had to be both efficient and affordable.

These requirements were met by adapting small, low-pressure air compressors for marine use, specifically to power a device that would become the favorite excavation tool of treasure salvors—the *airlift*. Similar in principle to the suction dredge, the airlift was a metal or plastic tube suspended in a near-vertical position underwater. A hose from the surface compressor introduced air into the tube a short distance from the lower, or intake, end. As the bubbles burst free within the tube, they expanded and rushed the length of the airlift tube creating a powerful upward water flow. Water was drawn into the intake end with sufficient velocity to carry with it the loose sediments nearby. A scuba-equipped operator positioned himself at the underwater intake of the airlift, guiding it over the sediments just as if he were directing a large industrial vacuum cleaner. Using the airlift, a salvor could now excavate—in a very short time—six, eight, or even ten feet into the loose sea-bottom sediments covering a wreck site. Correctly rigged, the airlift was a powerful tool, capable of picking up even an iron cannonball that could fit within the diameter of the tube.

The low-pressure compressors were also adapted for hooka use by addition of the necessary filters and an expansion tank. Many salvors developed their own multipurpose rigs using a slightly larger compressor capable of providing

CHAPTER 17: SALVAGE TECHNOLOGY 135

enough air for simultaneous hooka and airlift use. The small, inexpensive air systems gave the individual salvors an undersea working capability more thorough than any other salvors in history.

SEA-BOTTOM SCANNERS

Excavation and recovery, of course, were secondary considerations, for first the wreck sites had to be located. For lack of anything else, salvors for centuries had relied upon simple visual survey. In some present-day searches, divers are towed behind small boats and, in the clear water characteristic of much of the Caribbean region, can scan the sea bottom to depths sometimes exceeding sixty or even eighty feet. If a portion of an ancient wreck site were visible—perhaps cannon, anchors, or the elongated piles of ballast stone that mark many early sites—an *experienced* wreck hunter probably would spot it. Any exposed portion of a wreck will almost always be covered with a crust of coral or other sea growth. It will have exactly the same texture and color as the rest of the reef or sea bottom. Treasure salvors have trained themselves to detect the often subtle differences in shapes that might indicate a wreck site. Many Florida recreational divers have passed directly over large, plainly visible historic wreck sites and noticed nothing. Many wrecks are not visible at all, such as those in shoaling areas where deep, shifting sands have completely covered a site. A perfectly "normal" white sand or grassy bottom may hide a large and valuable wreck site.

Electronic technology has come to the rescue. The limitations of visual survey in wreck site location have been surpassed.

The first attempts of treasure salvors to detect the unseen on the reefs and sea bottom came with experimental underwater use of early metal detectors. Success was rare since most of the metal portions of a wreck site often were deeper than the limited range of the detectors.

The breakthrough came not with detectors, but with the improved versions of magnetometers which the U.S. Navy had first employed practically in World War II to locate underwater naval mines. The function of the magnetometers was to detect, measure, and record slight disturbances, or anomalies, in the normal magnetic field of the earth. Anomalies could be caused by any number of things, including the presence of a metal object capable of affecting the magnetic field. After the war, magnetometers were put to commercial use in such fields as geophysical prospecting for oil and mineral bodies and in the marine construction industry where the instruments were used to locate pipelines and cables. By 1960 the ponderous, bulky magnetometer of the war years had been replaced.

MODERN MAGS

The *mag* of today is a light, portable, and highly sensitive instrument. Many models were designed specifically for marine applications. Typical of advances in electronic technology, as the sophistication and efficiency increased, the price decreased to levels that now make some models of magnetometers available to nearly anyone. While the mags cannot detect the presence of gold, silver, copper, or brass, they do reveal the anomalies created by the presence of ferrous metals. In the world of the gold-seeking salvors, this translates specifically to anchors, cannon, and shipboard fittings that are a part of nearly every single ancient wreck site.

Modern magnetometers are as much a part of treasure salvage equipment today as is scuba equipment. Most units consist of a recorder the size of a suitcase, a long length of specially shielded cable, and a *fish,* or sensing head. In use, the fish is trailed behind the survey boat at a controlled depth and at a distance that insures it will be clear of the anomalies created by the boat itself. The actual survey conducted by treasure salvors in their search for sunken gold is hardly romantic or exciting. Rather, it may be infinitely boring and monotonous. With the mag zeroed into the specific regional magnetic flux, the survey boat with the fish in tow makes slow runs along evenly spaced parallel courses. Constant minor fluctuations in field readings are normal but the *hits*, those times the needle swings erratically to indicate a major anomaly, are not. At each hit, a weighted buoy is flung into the sea to mark the location. Additional runs may be made over the spot to confirm it. There is still no cause for excitement for each hit must now be checked visually by a diver. More times than not, the hoped-for, three-hundred-year-old wreckage will turn out to be a discarded oil drum, a fifty-year-old anchor chain, or an outboard motor someone dumped last month. Occasionally it may be a tantalizing bit, a cannonball or an ancient anchor. With luck it may be a ballast pile indicating the main portion of a wrecksite. Divers, responding to a mag hit, have plunged over the side to be absolutely startled by the sight of twenty or thirty huge cannon scattered haphazardly over a reef marking a shipwreck that has never been touched since the day it sank.

Modern metal detectors designed for fully submerged use are also used by salvors, not for wreck site location, but for assistance in excavation of a site, usually together with an airlift. Many wreck sites have been completely enveloped within the living coral of the reef itself. Explosives and pneumatic drills may be used to penetrate the reef, and many of the broken, cementlike blocks of coral may contain objects of great value, such as gold coins. A good detector, used underwater or on the surface, may save an enormous amount of time in evaluating each piece.

EXOTIC ELECTRONICS

Many modern treasure salvors, particularly those backed by companies and outside financing, employ even more exotic electronic locating instruments. Side-scan sonar has been used successfully to reveal the presence of unnatural sea-bottom shapes and formations possibly indicating wrecks. Sub-bottom profilers have also been employed to detect variations in densities of the sea bottom, such as those created by a wreck. Aerial visual survey has been used for years to scan large areas of remote reef lines. More recently, aerial techniques have included the use of infrared photography capable of differentiating sea-bottom features as deep as thirty feet.

WHAT PRICE GOLD-GETTING?

The cost of the new technological locating and excavating tools is not cheap in terms of dollar value, but is more than reasonable when one considers the value of the potential recoveries. It is unrealistic to think that a small group of salvors might outfit for treasure salvage with anything less than a ten-thousand-dollar investment. Compressed air systems, tanks, hoses, airlift components, and other essentials will certainly cost three thousand dollars. The price of a sensitive and reliable magnetometer starts as low as two thousand dollars. All of this gear will no no good unless there is a suitable boat to operate from. Anything larger than twenty feet with a reliable engine in used condition will cost five thousand dollars if it can be found.

An individual cannot and should not attempt to undertake this type of work alone, from the standpoints of both safety and efficiency. The ten-thousand-dollar investment may be split among three partners.

Keep in mind that one single gold coin may recoup the entire investment. And if one gold coin is recovered, chances are excellent there will be others. Whether or not gold is actually recovered from an ancient wreck site, it is very likely that other objects of silver, brass, glass, bronze, and other pieces valuable in their own right will also be found. No wonder gold seekers have turned to treasure salvage in such numbers that it has become a small industry in its own right.

Information is readily available for the hopeful treasure salvor. Two particular books are considered standards and include material on sunken wreck locations, techniques of research, search, and salvage, and wreck site and artifact identification. They are *The Treasure Diver's Guide,* by John S. Potter, Jr., and Robert Marx's *Shipwrecks of the Western Hemisphere.*

CHAPTER 17: SALVAGE TECHNOLOGY

No one will lead you by the hand to a treasure wreck, but the information and technology are available to launch your search in the right direction and possibly lead to the recovery of a fortune.

18
MODERN TREASURE SALVAGE

That the modern era of treasure salvage would have its beginning along the Florida Keys was only natural considering geographic, cultural, and economic factors of the region. During the colonial era, when Florida itself was a Spanish territory, the one-hundred-fifty-mile-long scythelike arc of low, reef-lined keys formed the north and east edge of the Gulf Stream route plied by the treasure galleons for two hundred fifty years. Even in the post-colonial period, the Straits of Florida were the natural route for all shipping between the Atlantic coast ports and the ports of the Gulf and Central America.

SALVORS OF THE KEYS

Thousands upon thousands of ships of every era and nationality sailed within scant miles of the coral reefs of the Keys. And every year for centuries, the Keys took their toll. In the 1900s it was no secret among Keys residents that the modern wrecks lay mingled with the bones of the treasure galleons not only along the reefs of the Keys, but even farther north along the Florida coast.

A few divers in hard-hat gear bumped along the bottom, casting curious eyes upon the ballast piles and heaps of ancient cannon. Some found the time to prowl around the old wreck sites and were surprised when a little digging turned up three-hundred-year-old artifacts and even a few objects of gold and silver.

When scuba became available, many Keys salvors put it to immediate use to take a better look at the ancient wrecks. These men were an independent and self-sufficient lot. They relied on fishing, lobster trapping, and, drawing upon their general wracking heritage, miscellaneous salvaging to earn a living from the sea. Each owned a boat or two, and all knew the surrounding reefs and sea like the palms of their hands. Treasure salvage became the right and second business of many Keys residents who found it could indeed be profitable. The first decade of what may be considered the modern treasure salvage era was free of publicity. The operations during those years were also free of any control or regulation from Tallahassee, the state capital, five hundred distant miles to the north.

Spanish gold in coins, bars, and jewelry was recovered during the late 1940s and early 1950s. Most of it was stashed away in salvors' closets or sold for very little more than the then-low bullion value of thirty-five dollars per ounce. Public awareness of the romantic aspects of treasure diving as well as appreciation of Spanish colonial history—factors that would have a major effect upon the value of recoveries—were still a decade or more into the future. While the ancient Spanish escudo coins were fascinating to look at, they were a bit of a departure from the more conventional and carefully struck coins commonly collected and catalogued by numismatists. Not only was the numismatic value of the coins minimal, but the gold bars with their tax and mint stamps and the personal jewelry were also considered little more than old gold and priced accordingly.

LAW OF THE SEA

With the absence of governmental control and regulation, a salvor's right to work any particular wreck site de-

CHAPTER 18: MODERN TREASURE SALVAGE

pended solely on his ability to hold it. The salvors worked secretly in small groups. Any publicity at all was a bad idea. Bragging about recoveries from a particular site often resulted in the competition anchoring over a man's wreck before he reached it in the morning. It's first come, first served in this business. The traditional and ancient law of salvage and the sea grants ownership to the man in control. The attitude of secrecy developed among treasure salvors holds true today more than ever. A standing rule of the trade is "if you found it, tell 'em you didn't, and if you didn't, tell 'em you did."

FROM SHADOW TO SPOTLIGHT

The quiet, shadowed world of the treasure salvors was suddenly dragged into the spotlight in the late 1950s when a major discovery was made near Fort Pierce, Florida. The now-famed Kip Wagner, a retired contractor, used a metal detector to uncover some silver Spanish pieces of eight on a beach. Those finds led to discovery of the source of the silver, a Spanish wreck only a short distance offshore. It was one of several treasure ships wrecked in a 1715 hurricane. When the size of the treasure trove became known, Real Eight, Inc., one of the first formal treasure salvage companies, was established. When the excitement finally died down, over $6 million in recoveries had been made, including some of the most remarkable gold artifacts ever taken from the sea.

The publicity was enormous. Newspapers and magazines devoted entire pages to the shipwrecks and sunken gold that might be found all over the Florida reefs. *Doubloons, escudos,* and *pieces of eight* became household words. When *National Geographic* magazine carried a major, color-illustrated article on the recoveries and the adventure, the existence of sunken Spanish gold and the reality of treasure salvage achieved almost instant credibility.

Legions of armchair adventurers followed the exploits of the salvors through the media. The more adventurous and curious sorts drifted down to Florida to check out things firsthand and perhaps to participate.

The avalanche of attention was both advantageous and detrimental to the treasure salvors. The worth of their recovered treasures began to increase sharply, for no longer were they selling simply old gold and silver. They were selling romance and history as well, commodities the media had made hot. But other interests began looking askance at the activities of the treasure salvors.

AFFAIRS OF STATE

First and foremost among these was the state of Florida, which, in 1962, came forth with a claim to 25 percent of all recoveries based upon the fact that the wrecks lay within the state territorial waters. To organize the treasure salvage operations, the state established a licensing and lease system that came down from Tallahassee as a maze of bureaucratic regulations.

This did nothing to endear the state to the independent salvors. They, to a man, believed that since they located and salvaged the wrecks, met all the expenses and took all the risks, they were damned well entitled to all the rewards, if indeed there were any. The state prevailed, however, and now the salvors were obliged to pay license fees and post substantial bonds before being granted individual eighteen-square-mile lease areas in which to search. Still other requirements had to be met before actual salvage operations could begin, at which time every salvor's private boat would have aboard an official representative of the state of Florida.

A bitter uproar resulted that was marked by daily exchanges of insults. The state claimed it stepped in to protect the irreplaceable historic sites that were the property of the people of the state of Florida and to stop the irresponsible looting by independent salvors intent only on fattening their own pockets. In reply, the salvors screamed that the state, never the least interested in the ancient shipwrecks until the multimillion-dollar recoveries came from the 1715 wrecks, was intent upon fattening its own pockets at the expense of the individual. They stated, rightfully, that the state had never previously located or salvaged a single wreck site. They

CHAPTER 18: MODERN TREASURE SALVAGE

considered its sudden interest just another example of the encroachment of government bureaucracy on individual enterprise. Even though the state salvage regulations had become law, so-called private, or illegal, salvage groups continued to operate up and down the Florida Keys. A few attempts were made to organize the salvors into a concerted front against the state, but the basic nature of the people involved made such cooperative efforts difficult. Some individual salvors' attempts to circumvent the state regulations wound up in court, to the delight of the newspapers. When the initial bitterness had subsided, the argument took on a more civil tone with the state preaching a line of idealistic archaeological concern and the salvors taking refuge in their right to free enterprise.

In 1969 the statement was made that "pirates and plunderers had taken tons of irreplaceable historical treasures," meaning illegal treasure salvors had succeeded in recovering substantial quantities of old gold and silver. This was true, but by the early 1970s the state had consolidated control over most of the salvors. Officials were hot on the heels of the few remaining pirates. About this time, the sharp rise in the price of gold brought forth new columns of treasure salvors who, rather than working under the state regulations, began looking toward other areas of the Caribbean.

PIRATES IN TROUBLED WATERS

They were quite right in their assumptions that, since Florida had been proven to have plenty of historic shipwrecks offshore, there would be even more along the thousands of miles of reefs that lined the ancient shipping routes elsewhere. The same factors that had made Florida's coast the first treasure wreck area to be explored and exploited assured that other potential treasure areas, namely the reefs of the Caribbean and Latin America, would be almost untouched. Access to these remote areas had been limited to natives, usually fishermen concerned with scraping out a marginal livelihood, not with purchasing expensive scuba gear and mags to pursue an activity that in no way guaranteed food on the table tomorrow. Many native fishermen knew of local wreck sites but had no means to work them.

Nearly all these wrecks rested in foreign waters. From the Pacific and Caribbean coasts of South and Central America to the West Indian archipelago, control of the waters was vested in separate governments that ranged from soverign nations to colonies and associated states, nearly all of which were able to establish their own laws governing their territorial sea and the shipwrecks therein. Most governments, however, remained unaware of the potential value of those ancient wrecks.

Treasure salvors saw the proverbial golden opportunity there and headed outward from Florida with their salvage boats, mags, and compressors. Most ignored the question of legality as they scoured the reefs near the ancient shipping routes and tried to track down the stories of wreck sites told by native fishermen. Many governments had no laws specifically regulating such activities as salvage, and those that did were usually incapable of effectively patrolling their waters and reefs. Some fortunes in gold and silver were taken from wreck sites in operations that were designed to get in, perform the salvage, and get out before the trouble started.

Many of the salvors found that excitement and adventure were not limited to the recovery of gold from old wreck sites. They were themselves considered pirates, but other pirates appeared, often in the form of corrupt local officials or competing salvage groups, using harassment and threats to get a piece of the action. Aboard the salvage boats, guns were as common as scuba tanks. The weapons were sometimes used for protection against unwelcome visits by coastal "traders"—the drug and arms smugglers—and armed fishermen who "fished" for anything of value in any way they thought might work. A few violent incidents occurred that resulted in the deaths of at least three salvors.

Salvage activity continued for over ten years. Several groups found their gold on the reefs off British Honduras (Belize), Nicaragua, and Colombia. The actual extent of the recoveries will never be known, but it certainly amounted to several thousand ounces of gold and much more silver.

CHAPTER 18: MODERN TREASURE SALVAGE

Spectacular finds were neither common nor easy to achieve. Some other groups met their expenses or managed a small profit; most went broke, sold what equipment they had left, and were lucky to be able to return home.

GOVERNMENT GUNBOATS

The nations of the Americas, particularly those of Latin America, became more nationalistic during the early 1970s. The governments took greater steps to protect what was theirs, including offshore reefs and shoals. Their primary concern was the protection of fishing grounds, but governments were also becoming aware of the numbers of pirate salvors recovering gold and silver from shipwrecks—gold and silver that was now referred to as national treasures and vital parts of the national heritage.

Illegal salvage was certainly not halted, but the increase in frequency of patrols gave the salvors something else to worry about. Nevertheless, most continued their operations under the noses of the gunboats.

Today most foreign governments and political states permit treasure salvage, but only under legal contract clearly spelling out the division of recoveries. Salvors are now usually permitted to retain half or three-quarters of all recoveries. But special clauses always attached allow the government to keep possession of items of special historical value or those that relate directly to the national heritage. In many nations contracts are now available only to the well-backed, larger salvage companies who can afford the expenses and "consideration payments" encountered in contract negotiation.

The Bahamas is one of the best areas to work for both companies and individual expeditions. The maze of islands features great expanses of reef line never fully explored, water and climate that permit year-round work, and a government extremely receptive to salvage proposals without the burden of payoffs or bribes. The Bahamas has approved as many as thirty salvage operations in separate lease areas at one time.

For every legal operation there is probably at least one

pirate group. Any salvor worth his salt knows that Bahamian patrol craft are stretched thin keeping track of drug and alien smugglers in their vast territorial sea.

Private treasure salvage has been banned by several nations, most notably Mexico. A government-authorized agency now strictly controls such activity. The Mexican government today retains complete possession of all recovered historical or archaeological materials.

Enforcement of the law was demonstrated clearly in 1974 when an illiterate Mexican fisherman discovered a Spanish wreck along the Gulf coast. He recovered a fortune in gold simply by free diving. The fisherman sold the gold to a local jeweler for a little more than bullion value; both wound up in a Mexican jail for violation of the antiquities laws.

Other Gulf of Mexico areas where private treasure salvage has been banned are Texas, where the wrecks of the 1565 treasure fleet were discovered a decade ago, and Louisiana, where in 1980 another great treasure discovery was made when a fishing boat snagged its nets in the wreckage of a sunken galleon.

TREASURE TODAY, TREASURE TOMORROW

The decade of the 1970s brought more changes, some good, some bad, to the treasure business. On the minus side was the increase in control and regulation of salvage operations, both within the United States and in the foreign Caribbean areas.

The same decade also saw the most dramatic gold price rise in history and a publicity boom about treasure salvors. The value of the gold treasure recoveries increased even more than the bullion value of gold, thanks to soaring numismatic and artifact demand. Both the romance of the Spanish colonial era and the adventure of treasure salvage itself have been translated to dollars and cents. Salvors who worked in the 1950s just shake their heads when they think of the minimal

CHAPTER 18: MODERN TREASURE SALVAGE

prices they received for artifacts in the early days of the business.

Today the pricing of artifacts, especially of gold in any form, is a matter of whatever the market will bear. It is a rare piece of Spanish or colonial gold that will not bring at least five times bullion value. One-ounce gold coins such as the English Rose guinea of the 1700s will bring about five thousand dollars each. Many Spanish eight-escudo coins in good condition will sell for double that. Many items are sold at auction. More are sold to buyers with cash in hand literally waiting at the dock.

More active treasure salvage groups than ever before are in existence today. The larger, well-known companies, because of the size of their operations, usually work on a completely legal basis. It is their recoveries that make the headlines. And for them the publicity is highly beneficial.

More adventurous are the loosely formed, smaller groups, most of whom outfit in Florida, then depart quietly for points unknown. Testimony of their existence, successes, and failures comes as newspaper headlines proclaiming that a group of treasure hunters has been jailed in some foreign port or with the sudden appearance of a shipment of gold or silver treasure items suddenly available for sale, most of which are quickly snapped up for cash by inside buyers.

The treasure salvage business, legal and illegal, is thriving today because it is no longer necessary, as it was in the 1950s, to make a grand discovery of a loaded galleon to recover a fortune. A mere handful of colonial period gold coins or three-hundred-year-old gold jewelry is a fortune in itself. Such a fortune might be found on any of thousands of historic shipwrecks.

19
ON REEFS OF GOLD

The eastern coast of Central America has always been one of the least developed regions of Latin America and the Caribbean. The varied, irregular coastline begins at Belize, just below Mexico's Yucatan Peninsula, where the second largest barrier reef in the world stretches over one hundred fifty miles to the south in a maze of tiny cays and shallow coral reefs. At Honduras the desolate "Mosquito Coast"—a series of false sand coasts protecting broad, shallow lagoons—runs east to Cape Gracios a Dios, then south along Nicaragua to Costa Rica and Panama, and finally eastward again to the Republic of Colombia in South America.

THE SPANISH MAIN TODAY

The entire Caribbean coast from the Yucatan Peninsula to Colombia is nearly eighteen hundred miles of sparsely populated mainland plains planted with citrus and bananas, quiet, sandy beaches lined with palms facing offshore reefs and islands, and the occasional presence of a ramshackle port town, noisy, dirty, enormously colorful. The port towns are typically Spanish, constructed about a central plaza. Beyond the sprawl of the outer shacks are the stone walls and rusting iron cannon of colonial fortresses. Vines and brush now cover the ruins of the massive walls where some of the cannon still point toward distant sea lanes once sailed by Cartagena treasure galleons bound for Havana. The rickety docks of the port towns are packed with every conceivable type of rundown boat. Boat owners and crews constantly seek any possible way to turn a buck or a peso.

This is not the Caribbean of the travel folders and cruise ships. Commercial tourism has yet to reach this remote stretch. The economy and the culture remain traditional and, it seems, fifty years behind the times.

Generally, the entire coast has retained the lively flavor of its wild history, and the inhabitants are certainly aware of their heritage. On some of the English speaking offshore islands, the surname *Morgan* is common. Each and every Morgan claims some sort of direct tie to Sir Francis himself, who three centuries ago took a little of the Spanish gold for his private coffers.

Tales and legends of buried pirate treasure abound along the Caribbean coast. If each held only a fraction of truth, the beaches would glitter with gold. The buried treasure stories are a bit exaggerated, but the stories of treasure wrecks on the reefs are not, for the Caribbean coast of Central America is a graveyard crowded with colonial shipwrecks.

THE BUSINESS OF TREASURE

Treasure salvors have always considered this coast and the off-lying reefs to be wide open for operations. During the 1960s and 1970s as many as three or four treasure salvage groups operated simultaneously. Each group solemnly claimed to be backed by legal authorization of the appropriate government. Usually the claim was supported by a letter in Spanish signed by some unheard-of minister. Such letters were provided for unmentioned consideration by almost any minister, the only person in his government who knew such a document existed. A few valid contracts were issued now and then, but their legality and longevity were

CHAPTER 19: ON REEFS OF GOLD

too often dependent upon the frequent government overthrows that characterize Central American politics. The de facto versions sometimes proved more practical in actual use, as long as the minister and certain indispensable local officials remained confident they would be taken care of if and when recoveries were made.

The treasure salvage groups, most of whom were comprised of United States citizens, did not merely search the reefs and leave. They often took up long-term residence, accepted as part of the local waterfront scene in the port towns and on some islands. In dress, living quarters, and equipment, the serious groups were wisely unpretentious, taking care to bring no unnecessary attention to themselves or their operations. All dabbled in mechanical repair, lobster diving, and light commercial salvage, doing favors for locals and making a little money on the side.

The salvage groups headquartered themselves in different towns and usually different countries, each reigning over its own territory. Contact between separate salvage teams was infrequent. The rare meetings called for equipment or boat bartering were marked by mutual caution and distrust. The relationships between the salvors of the Central American mainland coast were remarkably similar to those between the wrackers of the 1800s.

PRIME HUNTING GROUND

The treasure salvors concentrated their efforts on the reefs and shoals near the mainland coast, but all, at one time or another, ventured to the distant offshore reefs over one hundred miles east of Nicaragua. Of all the shipwreck hotspots in the Caribbean, these reefs are probably where the next major recoveries will be made.

The four remote, unpopulated reef groups—Serrana, Serranilla, Quito Sueno, and Roncador—are fringed with a collective one hundred twenty miles of deadly coral reefs once directly in the path of Havana-bound treasure galleons. All the reefs are roughly similar in size and other characteristics.

Serrana Bank has the largest cay, a half-mile-long desert island with glistening white sand dunes rising to a height of almost thirty feet, a few palm trees, and a densely brushy interior where thousands of sea birds nest. The island and reef are incredibly beautiful and similarly dangerous. There is no protection from storms, and the nearest truly safe harbor is at least a full day's voyage away. The reefs' infrequent visitors include smugglers and the wide-ranging fishermen in search of lobster and turtle, all of whom will be more trouble than help to the salvors. Little has been written about this forgotten corner of the Caribbean. The best picture is presented by Peter Matthiessen in his book about the Caribbean turtlers, *Far Tortuga*.

The tale of these reefs is as captivating as their pristine beauty. They were probably discovered by shipwreck; in fact, their first accidental resident, Pedro Serrana, was the sole survivor of a Spanish ship lost in 1523. Serrano was half dead when he crawled ashore on the tiny cay that would later bear his name. He existed on bird eggs and turtle blood in a classic struggle for survival that some scholars believe may have inspired Defoe's *Robinson Crusoe*. After five years alone Serrano was joined by another shipwrecked Spanish sailor. Rescue finally came in the form of a passing ship that answered their smoke signal, eight long years after Serrano had crawled ashore. Serrano was eventually paraded before the courts of Europe as a living symbol of both the adventure and the hardship that would be found in the strange New World.

Only a few years after his rescue, the annual procession of treasure ships bearing the Incan gold taken by Pizarro passed close by Serrano's namesake reef. Whether any of those wrecked is uncertain. A century later, however, a major galleon of the 1633 treasure fleet was documented as lost near Serrana. Aboard was a cargo of nearly ten thousand ounces of Colombian and Peruvian gold along with tons of silver. No salvage has ever been recorded.

Serrana Bank and the adjacent reefs continued to trap ships long after the last of the treasure galleons of Spain was gone from the seas. By the 1850s the tiny islands were known as a rich source of *guano,* a valuable nitrogen-rich

CHAPTER 19: ON REEFS OF GOLD

fertilizer much in demand in the United States. During those years of Manifest Destiny, it is not surprising that the islands were claimed under the Guano Act of 1856. Swarms of American schooners visited the islands to mine the deposits. When factory-produced fertilizers replaced guano, the islands lost their economic importance. The United States then passed the reefs back to the other claimants which included Honduras, Panama, Colombia, and Nicaragua.

Even as the era of sail ended, the reefs continued to take their toll on shipping. Today the reefs are visible from many miles away because of the towering rusted skeletons of several four-hundred-foot steel freighters. They, like the ancient galleons before them, met their end on the deadly coral reefs.

Although a joint claim still technically exists regarding ownership of the reefs, they are controlled by the Republic of Colombia. In reality, the reefs are no man's land. The infrequent patrolling visits of the Colombian gunboats discourage neither foreign fishermen nor treasure salvors. Some excellent recoveries have been made by treasure salvors, but the majority of lost treasure still waits on the reefs—probably as much as $10 million in bullion value of gold alone, not counting numismatic and artifact worth. Treasure salvors still operate on the reefs, and exactly what they recover is known only to them.

In 1976 a run-down shrimp boat converted to salvage use anchored in the open harbor of Georgetown, Grand Cayman. Cayman has served treasure salvors in the past, not only as a convenient geographical stopover, but also as a very convenient financial facility offering secrecy, substantial tax benefits, banking services, and trans-shipment services. Some eighty pounds of Spanish gold purportedly salvaged from a wreck site on Quito Sueno were offloaded and disposed of in financial channels. The total worth of the recovery was several million dollars. Not a bad two months' work for six men and a secondhand shrimp boat, eh?

JUMPING OFF FROM KEY WEST

If the western Caribbean reefs are a bit far from home for a prospective treasure salvor, he or she may still experience some of the excitement right in Florida. The Florida Keys, where modern treasure salvage got its start, remain a center of salvage activity. Key West, the flamboyant tourist town at the end of the Keys, headquarters several salvage groups. The local history of rumrunning and wracking, the distinctive Bahamian architecture reminiscent of the 1800s, and a good collection of offbeat bars catering to shrimpers, lobster fishermen, and tourists form a natural background for this unusual and often controversial business.

The bars along Duval Street and in the Old Town section provide convenient meeting places where divers may hire on, deals are made, boats and equipment are bought and sold, traded or borrowed, news is exchanged, and recoveries are often sought and sold. Common occurrences among the salvage companies are temporary capital shortages. Things tend to be either very good or very bad at any one time. In lieu of cash divers sometimes are paid in eight-reale silver coins, which, adding to the uniqueness of Key West, are often sold in the street.

MEL FISHER, SUCCESSFUL SALVOR

Active salvage is conducted by a number of groups up and down the Keys, but Key West remains the center of activity thanks largely to the presence of Mel Fisher, perhaps the most accomplished of all treasure salvors. Fisher, after a long search of many years, located the scattered remains of two galleons in the 1622 treasure fleet, the *Atocha* and the *Marguerita,* about thirty-five miles west of Key West.

International notoriety came to Fisher on several counts. First was his determination to locate and salvage the wrecks in the face of great technical problems and personal tragedy. Following that came his well-publicized battle of many years in state and federal courts to win control of his wrecks and recoveries. Finally, on the bottom line, came his success. Fisher has now recovered many millions of dollars in Spanish gold coins, bars, and jewelry.

Fisher's wreck sites were extremely scattered with the trails of wreckage many miles long and covered by over

CHAPTER 19: ON REEFS OF GOLD

twenty feet of sand in places. Years earlier Fisher had developed an excavating tool widely used by others since. Referred to variously as *propblasters, mailboxes,* and *blowers,* they are simply steel, elbow-shaped tubes designed to fit over a ship's propellers. The propblaster directs the wash downward in a powerful stream capable of eroding away the overburden covering a wreck site. Fisher had always used relatively small units, but another Key West salvor, Dennis Standefer (president of Seaborne Ventures, Inc., a Key West salvage company) designed and built enormous units that fit over the eight-foot-diameter propellers of his large salvage ship, the *R/V James Bay.*

In a rare effort where two independent and separate salvors worked together successfully, Standefer put his huge blasters over Fisher's wreck site in the summer of 1979. Several uneventful and disappointing weeks of work followed in which divers explored hundreds of barren excavation holes in the deep sand.

Encouraging bits of wreckage appeared, then, finally, the sought-after gold. For nearly one week the Spanish gold hidden by the shifting sand for more than three hundred fifty years was brought to the deck. Most of it was found and recovered by a twenty-nine-year-old ex-painting contractor recently turned treasure salvor, Rich Banko. The gold came up as delicate bracelets, a spectacular ten-foot-long money chain gleaming as brightly as the day it was made and worth over a quarter-million dollars, and many stamped bars. The booty represented well over a million dollars.

What they found was only a small part of the lost gold aboard the galleons. The rest is waiting to be recovered—and will be.

Many first-time salvors hire on with established companies to pick up experience, then set out on their own, legally or otherwise, with their own companies or in informal groups. Either way the prospects seem favorable, and the treasure salvage business should continue to grow in the foreseeable future. Both the demand for and value of recoveries will increase as further advances in technology lead to the discovery of even more treasure wreck sites.

Only a fraction of the sunken gold in the Americas has been recovered, and that which is left on the reefs is worth more now than the ancient Spaniards ever dreamed. And those able to recover it will make fortunes.

20
EAST COAST COIN BEACHES

There's another angle to beachcombing besides shells and seaweed. Individual gold seekers have excellent chances of success all along the U.S. shoreline. Some of the most steadily productive *coin beaches* are those along the mid-Atlantic coast, areas within two hours' driving time of millions of people. The sources of the many coins are the unusually heavy concentrations of colonial and post-colonial period shipwrecks buried on the offshore shoals.

GRAVEYARD OF THE ATLANTIC

The section of coast from Sandy Hook, New Jersey, south to Cape Charles, Virginia, a distance of roughly two hundred miles, has long been one of the maritime graveyards of the Atlantic. Its notoriety may be attributed basically to a combination of two factors: first was the extremely heavy regional shipping traffic serving the busy ports of New York, Delaware, and Chesapeake Bay; second was the onshore gale winds of the winter nor'easters driving many of these ships to destruction on the shoals and barrier beaches.

Mid-Atlantic maritime records note not only the passage

of warships and countless merchantmen over three centuries, but also the activities of a legion of pirates and wartime privateers in the 1700s who found the many small bays and barrier beaches perfect for the operations.

New wrecks are few. Modern navigation and reliable power see that the shoals claim a ship only on rare occasions. But the remains of more than two thousand shipwrecks already lie along the barrier beaches and islands of the coast, evidenced most often by the great, broken timbers that wash ashore after the northeast storms have eroded shoals and beaches. For well over a century these beaches have yielded thousands of historic coins once lost with the wrecks. Many of the coins are gold.

After nor'easters, recoveries of single coins are not at all uncommon. Larger finds in recent decades amount to true fortunes at today's prices. Construction work in 1937 near a beach at Manasquan, New Jersey, turned up a cache of sixty-five United States gold coins dating to the early 1800s. The finder, a numismatist and local historian, attributed the cache to a retired sea captain and estimated the numismatic value of his find at "about eight thousand dollars."

A decade later, in April 1948, twenty-six one-ounce gold Portuguese *Johannas* were dug from the sands of a beach at Highlands, New Jersey. Since numismatic interest in foreign gold coins was still lacking, the value assigned to these mid-1700s mint-condition gold coins was fifty to one hundred dollars each.

Both recoveries were made without the aid of metal detectors, and, as one would expect, each resulted in a minor gold rush as hundreds of people armed with rakes and shovels dug up the nearby beaches. At the time both recoveries made exciting news but not much more; today, gold and numismatic values would make such finds worth a fortune. The Portuguese gold Johannas would be worth about one hundred twenty-five thousand dollars, and the sixty-five United States gold coins in excess of two hundred thousand dollars.

What may be the most steadily productive beach in the United States for historic gold, silver, and copper coins is located about one-half-mile north of the Indian River Inlet bridge on Delaware's Atlantic coast. The source of most of these coins seems to be the wreck of an English merchantman lost on the shoals in 1785. One or two other wrecks from the same time period in the immediate area may also contribute. Thousands of copper halfpennies from the mid-1700s have been found here but, as always, it is the gold coins that command attention. Accurate estimation of the number of gold coins found here is impossible, but it would be safe to say the total is well over one hundred. Most of the coins are gold English Rose guineas containing about one ounce of gold and bearing dates from the mid-1700s, together with a few Spanish coins.

"CLAMMING" FOR GOLD

In autumn of 1980 I joined a salvage team anchoring several hundred yards off the Indian River beach in Delaware. While gold seekers had always done well on the beach, prior to that time no gold coins had ever been recovered directly from the sea. As wrecksites along shoaling coasts are almost certainly buried and often scattered, an industrial type of operation using a clam bucket was employed. Working from the deck of a steel barge, a heavy crane raised and lowered the bucket to the sea bottom fifty feet below. Each retrieval of the bucket brought to the surface a ton of bottom sediment that was dumped into a screened hopper on deck. Hoses washed away the silt, sand, and mud. Left behind was a layer of shell and glass fragments, pieces of conglomerate, congealed chunks of tar and sand, bits of corroded iron, and, in testimony to the enduring efforts of surf fishermen, snarls of monofilament line, sinkers, and rusted hooks.

In every load of washed sediment were a few old copper halfpennies, each cloaked in a green-black layer of corrosion that told of centuries of immersion in the sea. The copper coins confirmed the presence of the scattered 1785 wrecksite. By late morning a plastic bowl held well over a hundred halfpennies. While neither the location, Delaware, nor the method, *clamming,* reflected much of the popular image of treasure salvage, what happened next did.

Just before noon, the gloved hands of crewmen sorted through another load of bottom sediment to recover still more halfpennies and a few two-hundred-year-old shoe buckles. Then, from the jumble of shells and gravel, came the unmistakable glint of gold. Excited voices joined the din of the gulls and the whine of deck machinery as the crew crowded together, reveling in the backslapping and handshaking that have accompanied the recovery of gold from time eternal.

GOLD GUINEAS FROM KING GEORGE

The coin was in perfect condition, as lustrous and bright as the day it was made. The Delaware sunlight played off the still-wet features of the Rose guinea, the stylized bust of King George III with long flowing locks, the elaborate English coat of arms, and the Latin words telling of a different time: *George III By The Grace Of God King Of England ... 1766.* In the work that followed, more golden guineas were found in the shells and gravels of the sea bottom, these bearing dates of 1752, 1758, and 1759. All were in mint condition with the exception of one with a small scratch.

The real value of such recoveries may be determined only in an actual sale situation. Those Rose guineas, with their one ounce of gold, brought five thousand dollars each. The recovery made that day from the barge confirmed that hundreds, perhaps thousands, of these gold coins will in time be washed up by the storms on the beaches. After winter storms, there are a handful of gold seekers who regularly work the beaches. They use modern detectors costing several hundred dollars, and they are looking for—and finding—historic gold coins worth thousands each.

The original image of the professional graverobber, or huaquero, carried none of the evil and unethical connotations attached to the occupation today. Huaqueros were then working men trying to earn a living from the earth much in the way of a miner, the sole difference being they took their gold from the graves and not from the gravels.

IV
FROM
THE
GRAVES

21
THE GOLD OF EL DORADO

The Indians, named in ignorance by Columbus, mined and worked gold before the time of Christ, a deity about whom they knew nothing and could have cared less, having enough of their own gods to worship. They venerated gold to such a degree that the metal was never degraded to common use as a monetary medium and only rarely as a symbol of material wealth. Gold was deemed so noble by the Indians that only the finest art and the most sacred religious expression and sacrifice justified its use. As goldworkers the Indians, particularly those of what is now Colombia, achieved the highest level of craftsmanship and technical competence, in many cases exceeding that of Old World cultures.

LUST FOR A LEGEND

When the Spanish first drove their way into the jungles and mountains of Central and South America, their sensitivities were still burdened by medieval limitations. Their priorities for the magnificent golden artwork they acquired were first to melt it down into bars and second to find its source, that vague place of infinite wealth manifested in the name *El Dorado*.

The legend of El Dorado began as early as 1510, rapidly embellished as the Spanish heard ever more Indian tales of a spectacular ceremony conducted on a remote mountain lake deep in the unexplored interior of Colombia. Every piece of worked gold the Spanish acquired in trade, captured, or dug from graves served to reinforce the legend. To the growing ranks of would-be *conquistadores,* El Dorado came to mean nothing less than the source of all the gold in the Americas.

The details of the first accounts to reach receptive Spanish ears have been lost in time, for conquistador and journalist alike tended to exaggerate anything to do with gold. In enticing stories, the Indians told of a coronation-like ceremony inaugurating the reign of a new god-ruler. The ceremony concluded with the massive sacrifice of golden objects cast into the bottom of the deep lake.

When the Spanish first arrived on the shores of South America, Indians were still alive who had personally witnessed the last of these great golden ceremonies. Those accounts were recorded by one of the gold-seeking Spaniards, Juan Rodriguez Freile, who wrote:

> At this time they stripped the heir to his skin and annointed him with a sticky earth upon which they placed gold dust so that he was completely covered with this metal. They placed him on the raft, on which he remained motionless, and at his feet they placed a great heap of gold and emeralds for him to offer to his god. On the raft with him went four principal subject chiefs, decked in plumes, crowns, bracelets, pendants, and earrings, all of gold. They, too, were naked, and each one carried his offering. As the raft left the shore, the music began, with trumpets, flutes, and other instruments, and with singing that shook the mountains and the valleys. ... The gilded Indian then made his offering, throwing out all the pile of gold into the middle of the lake, and the chiefs who accompanied him did the same.... From this ceremony came the cele-

brated name of El Dorado, which has cost so many lives and fortunes.

Freile's account was recorded in 1633, more than a century after the Spaniards launched their search for El Dorado, but at a time the quest was still conducted with unabated determination. It was then believed that the origin of the ceremonies was a "golden god" who in prehistoric times had plunged from the sky into the lake. Scholars and historians now consider the possibility that the "god" was a meteor. While the origin is open to conjecture, there is no question about the validity of the legend itself, for the actual existence of golden sacrifices has been confirmed by recovery.

GUATAVITA'S GOLD

The search for El Dorado covered much of the northern tier of South America but, even from its beginning, has traditionally centered upon Lake Guatavita, a mountain lake about thirty-five miles northeast of the present city of Bogotá at an elevation of eighty-five hundred feet.

The first of many imaginative attempts to recover Guatavita's gold was made in 1545 by Hernan Perez de Quesada. Quesada pressed hundreds of Indian slaves into bucket brigade service. After three steady months of hauling, they managed to lower the water level of the lake by about ten feet. From the newly exposed lakebed near the shoreline, Quesada is reported to have recovered several thousand ounces of worked gold, an encouraging treasure, but small in comparison with the gold the Spanish had already captured from the Indians or looted from their graves.

The next major stab at Guatavita's gold was made in 1580 by Antonio de Sepulveda, a wealthy Bogotá merchant. With an enormous work force—eight thousand Indians—Sepulveda attempted to drain the entire lake by digging a great notch in the natural earthen bank around the crater lake. The torrent of rushing water that suddenly escaped undercut the walls of the excavation, triggering a landslide that killed many Indian workers. But the assault on the lake paid off as the water level dropped sixty-five feet. Sepulveda holds the record at Guatavita with his great pile of treasure—golden breastplates, staffs, figures of eagles and serpents, and an emerald the size of an egg. His recoveries again represented only a very small part of the gold believed to be at the bottom of the lake, but they brought further credence to the legend of El Dorado and served to inspire gold seekers for centuries to come.

DIGGING AND DRAGGING TO NO AVAIL

Over the years a number of lesser and totally unsuccessful efforts were made by Spaniards licensed by the crown to recover the golden treasure. The passage of time brought about the end of the Spanish colonial empire but did nothing to diminish the lure of Guatavita's gold or the legend of El Dorado. Even as the last of the colonial Spanish departed in 1820, another attempt was made, this one widening Sepulveda's original excavation. The great amount of work and money was expended in failure, for the steep gradient of the walls could not be controlled. Landslide after landslide put an end to the project.

The next full-scale attempt (1899), a truly innovative and very expensive effort, required the construction of a long tunnel beneath the lake coming up into the lake bed itself to drain the water completely. This bizarre project actually succeeded—but only in draining the lake. For the first time ever, Guatavita was dry, but the centuries of mud and slime that formed the lake bed quickly dried and baked to a bricklike consistency that prevented effective digging and sealed up the artificial drainage system. Before the baked mud could be penetrated and searched, or any significant amount of gold recovered, Guatavita filled back up to its original level.

In 1911 a well-financed British expedition employed a large steam shovel mounted on a barge that was floated to the center of the lake. The great bucket groped blindly in the depths for several months. A few encouraging bits were recovered but, in the end, the steam shovel approach, like all the others, proved unequal to the gold of Guatavita. As the twentieth century moved along, an assortment of modern contrivances including enormous pumps, draglines, divers, airlifts, and dredges were brought to bear, none successfully. Powerful explosive charges were used in drainage attempts, also unsuccessful.

Forays by treasure hunters continued into the early 1960s when Lake Guatavita still represented El Dorado and one of the greatest recoverable golden treasures on earth. All attempts on Guatavita's gold by individual private gold seekers were finally banned in 1965 when the Colombian government enacted a law making the lake a major part of the national cultural and historical heritage.

22
GOLDEN TOUCHES

The worked gold ceremoniously cast into Lake Guatavita was representative of the creativity and craftsmanship among a number of Indian cultures from Mexico to southern Peru. Gold had been mined and worked by the Indians since 2,000 B.C., and the craft reached its peak about the time the Spanish arrived. The less developed tribes were limited in their goldworking ability to the production of simple hammered ornaments. It was among groups with a high degree of cultural complexity and specialization of labor that goldworking techniques became truly advanced. In these societies the goldsmiths enjoyed a particularly honored position and were free to exercise their growing creativity with gold.

INDIAN GOLDSMITHS

Several subcultures of the Aztec and Incan civilizations produced notable work, but it was the Indian societies of Colombia that truly excelled in gold craftsmanship. By the time the Spanish put an end to their technological and artistic advancement, Colombian Indian goldsmiths had mastered every single technique of goldworking known to Europeans. Their level of development was such that, with the single exception of electroplating, their techniques included all those used by modern goldsmiths.

Indian goldworking originated in Peru and spread northward, reaching Colombia about 700 B.C. and Panama and Costa Rica about A.D. 300. Advanced goldworking techniques were not carried into Mexico until comparatively recently, about A.D. 900.

RICHES FOR THE AFTERWORLD

Most of the worked gold in the Aztec and Incan civilizations was used for religious purposes and, as such, lasted for centuries. In the Indian cultures of what is now Costa Rica, Panama, Colombia, and Ecuador, most of the worked gold eventually accompanied the dead into their graves.

The gold committed to the graves was in every form the Indians ever fashioned: helmets, goblets, breastplates, figurines, nose rings and earrings, crowns, pendants, beads, and even bells and trumpets. Scholars still debate the true purpose of the burial gold. Some believe it was a symbol of the station or status which the individual had achieved in life, others believe it served a more functional purpose, specifically for use in an afterlife. Another idea is that the more gold an Indian carried with him on his mysterious journey of death, the more esteemed he would be when he arrived at his equally mysterious destination. Whatever the true purpose, an enormous quantity of worked gold went into the graves.

MINING AND REFINING

To provide quantities of raw metal for the goldsmiths, a complex system of mining was set up to exploit the rich deposits of the northern Andes. Most of the gold came from placers where slaves worked the banks and gravels with wooden implements, separating and recovering the gold in large wooden trays similar to bateas. Before the arrival of the Spanish, some placers were so rich that the Indians simply waited for the seasonal floods to subside, then searched the ground for visible nuggets.

The Indians also acquired a much smaller part of their gold from lode sources. The miners reached the veins deep in solid rock through twisting tunnels only three feet in diameter often declined at angles of forty degrees or more. Tunnels have been found as long as eighty feet, remarkable achievements when one remembers the Indians had no iron implements at all. Because of the great physical effort required to construct the tunnels in the hard rock, they were kept as narrow as possible without even the room to allow a single worker to turn around in. The extracted ore was then crushed and washed to recover the gold.

A crude smelting process was performed in ceramic furnaces. Blowpipes generated sufficient heat to separate the metal from fine sands and particles of crushed ore. The smelted gold contained any number of other metals such as silver, iron, and copper, all of which modified the working characteristics. To achieve the desired purity, an ingenious refining process was devised. The gold particles were heated to a dull red color while being mixed thoroughly with common salt (sodium chloride). The undesirable elements would then form metal chlorides which would vaporize and be driven off, leaving gold of a fairly high purity.

ANDEAN GOLDWORKING TECHNIQUES

Hammering was the most primitive and first-developed technique for working gold. Although a simple process, it required a considerable degree of skill on the part of the craftsman. Long, laborious hammering was required to fashion the thin, broad gold sheets favored by many cultures for breastplates and funerary vestments. After extensive hammering the spreading gold sheets tended to become brittle and subject to cracking. To prevent this the *annealing,* or tempering, process was used. The gold was heated, then cooled in water before further hammering. Repetition of hammering and annealing eventually resulted in large, smooth sheets of uniform thickness.

The hammered sheets of gold offered the craftsman many creative directions. Decorative designs were traced on the sheet and pressed outward from behind. The raised outline could be further accentuated by additional work on the front with chisels and punches. All fashioning of sheet gold was accompanied by continuous annealing to maintain the maximum workability of the metal.

Several simple joining techniques were used to create hollow objects from flat sheets, such as fastening with golden nails and staples or hammering a double-folded seam. *Fusion welding,* a much more complex and delicate technique, was used only on the finest pieces. Two sheets of gold were carefully heated to a point just below the melting temperature. Addition of small amounts of copper acetate, obtained from the action of vinegar (acetic acid) on metallic copper, brought about an actual molecular fusion. This true weld was very strong and artistically perfect, nearly invisible to the eye.

Another goldworking technique was the lost wax casting process in which several Colombian cultures became particularly proficient. Carvings of the finest intricacy were created in easily workable and moldable wax. That image then was transposed into a ceramic material for the final mold into which the molten gold was poured to set. Virtually any shape could be reproduced accurately in gold. The artistic effectiveness of lost wax casting is reflected in the exquisite statuettes and figurines that represent the highest level of the Indian goldworkers' development.

Much of the gold used in Indian metalworking in Panama and Colombia was actually *tumbaga,* a two-part copper and one-part gold alloy. Tumbaga offered two important advantages to goldsmiths apart from its pleasing rich reddish color. The alloy melted at only about eight hundred degrees Centigrade, two hundred fifty degrees lower than pure gold, making both the smelting and casting operations in the small, crude furnaces a good deal easier. The formative qualities of the alloy were better than either copper or gold, allowing even the most delicate designs in a mold to be captured in the casting.

CHAPTER 22: GOLDEN TOUCHES

Gilding was yet another technique the Indians applied to their worked gold, especially to the tumbaga pieces where it created a surface appearance of pure gold. The finished artwork was heated until it glowed red, turning the surface copper of the alloy to copper oxide. The copper oxide was removed in an acid bath leaving essentially pure gold with its characteristic yellow luster on the surface.

A second gilding technique involved treating the surface of the artwork with a strong oxidizing agent of mineral origin. Iron sulfide, for example, eliminated the copper. Repeated treatments resulted in a solid gold layer on the surface.

Gilding was applied to objects of special significance as well as those intended for long use. The pure gold surface prevented the oxidation and discoloration that eventually affected ungilded tumbaga. For centuries people erroneously believed that the Indians favored gilding simply to economize on gold. Many Indians were killed by the infuriated Spanish who believed that the gilded goldwork received in tribute was an attempt to deceive them with a lesser alloy.

LOST IN JUNGLE GRAVES

Some Colombian Indian goldworking societies produced no raw gold themselves. They relied on an organized, wide-ranging trading system for their needs. The gold artwork was also traded and circulated hundreds of miles from its point of origin. However and wherever it was worked, the majority of gold ever fashioned by the Indians was given to the gods, as at Guatavita, or buried with the dead.

The legacy of the Indian goldworkers—their exquisite, priceless, and irreplaceable golden artwork—remains lost but hardly forgotten in thousands of jungle graves throughout Central and South America.

23
PROFESSIONAL GRAVEROBBERS

When it came to matters of gold, the values of the Spanish opposed those of the Indians. The differences caused the latter unimaginable suffering.

GRAVEROBBING FOR GOLD

First among the Indian sanctuaries to be desecrated were the sacred graves of their dead, which, to the Spanish, were nothing more than convenient repositories for the gold they sought. Graverobbing was nothing new to man. Egyptian tombs had been subjected to looting for thousands of years before the Spanish ever heard of El Dorado. Columbus's men probably engaged in the practice during his last voyages, but not to any great extent as the admiral's main concern rested in discovering the true Indies, or at least the passage leading to them. By the time Balboa crossed the Isthmus of Panama to discover the Pacific Ocean in 1513, graverobbing had become the major preoccupation of his soldiers. The men considered this wonderful opportunity to enrich their own pockets an earned right. Cortes's troops pursued the same lucrative pastime during their long march to the Aztec capital in Mexico. The Spanish worked the graves of Central America at their leisure. Even after the rich placers of interior Colombia were discovered, the graves were considered to be a better, faster source of gold.

Official Spanish reports show that in 1533 and 1534, the first two years of Spanish presence in the Colombian interior, the graves yielded 110 pounds of pure gold and 53 pounds of tumbaga pieces, which they referred to as "base gold" to indicate both its impurity as well as their disdain for the alloy. The following three years proved increasingly productive as more Spanish arrived to scour the jungles for graves. In those years digging produced 545 pounds of pure gold and 176 pounds of tumbaga. The Spanish tallied their loot from the graves as mere weight, although it was recovered in the form of thousands upon thousands of pieces of spectacular artwork, as it was all destined for the melting furnaces anyway. The figures given are official, that is, the crown was successful in exacting its tax. In reality, the weight of gold taken from the graves was three or four times higher, as most of it became the personal fortune of the man who dug it up. The Colombian graves proved so rich, and the subsequent private enterprise so rampant, that the Spanish Crown decreed in 1537 that "you shall not take gold from the sepulchres," but not out of respect for the sanctity of the graves. For the edict closed with "except in the presence of the Overseer, Officials of His Majesty, or their Lieutenants." The clause assured that the crown would not be cheated out of its share of the graverobbers' spoils.

DEALING WITH THE DEAD

The Spanish exploitation of the Indian graves continued for the duration of their New World empire, and, when independence came to Latin America, the graverobbing tradition was left in the capable hands of the *huaqueros* (literally, "tombers"). The term *huaquero* was a Spanish derivation of the Quechua Incan word meaning one who digs *huacas,* or graves, to recover *huacos,* the sacred objects buried with the dead.

CHAPTER 23: PROFESSIONAL GRAVEROBBERS

The original image of the professional graverobber, or huaquero, carried none of the evil and unethical connotations attached to the occupation today. Huaqueros were then working men trying to earn a living from the earth much in the way of a miner, the sole difference being they took their gold from the graves and not from the gravels. If anything, they were pitied because their profession required them to disturb the dead and thus risk the possible superstitious consequences. The most widely believed of superstitions was that the ghost or soul of the Indian might follow the unlucky huaquero for the rest of his days and, upon his death, mark his own grave. The Indian spirits were believed to gain revenge by disturbing the huaquero's bones from their eternal sleep.

NEW PROFITS FROM OLD GRAVES

Superstitions or not, graverobbing became a very profitable business. In some cases it was a family tradition with diggers tramping the jungles of Costa Rica, Panama, Colombia, and the adjacent nations wherever graves held the promise of gold. Great quantities of magnificent golden artwork were recovered and sold for the prevailing gold bullion price of about twenty dollars per ounce.

It did not matter that this gold was an irreplaceable treasure, representing the essence of a violently terminated society. Archaeological considerations were still a century away, and the artwork was melted down into more convenient forms. An enormous amount of other cultural material, including turquoise, bone, ceramics, and jade, was also uncovered in the graves. Compared to gold, this material was considered worthless and discarded.

Like the Spanish, the huaqueros were a cultural and archaeological tragedy, but a phenomenal economic success. The highly profitable profession was not at all limited to anonymous peasants.

Francisco Morazán, a leading Central American political figure, turned to graverobbing in Panama in an attempt to recoup his lost personal fortunes. Morazán hired hundreds of diggers in one of the first organized mass exploitations of the graves. Some years later, in 1859, so many rich, untouched graves were discovered in Panama's Chiriqui region that a minor gold rush developed. In only two weeks, over one thousand pounds of gold was dug from the graves. Even today several of Panama's wealthiest families trace the start of their fortunes directly back to their ancestors of a century ago who walked out of the Panamanian jungles covered with mud, their arms bearing the sacks of pre-Colombian gold artwork dug from the graves.

MYSTERY AND ARCHAEOLOGY

Much of the responsibility for the mystery shrouding the pre-Colombian Indian cultures of Central and South America must be borne by the huaqueros. Much of the grave material available for study today was collected and preserved when the embryonic science of archaeology made its presence known in the jungles about 1900. In those days, the budding science lacked much of the pure intention it is credited with today.

Archaeologists who ventured into the jungles were of two kinds: some were bona fide scientists representing interests of museums and universities in the United States and Europe; others merely assumed the title for ease of operation while going about their main concerns of adventurism and profit. Gold continued to be taken from the graves, a few pieces reaching the security of museums, the remainder going into the melting pot. Even as late as the early 1900s, the huaqueros stumbled on sites that were priceless, both archaeologically and economically. They recovered the gold and left the remainder scattered in a manner that was scientifically worthless for later study. Among these sites was a twenty-thousand-square-foot area which, from the charcoal and fragmented crucibles present, was judged to be an Indian foundry used by goldsmiths in pre-Colombian times, probably the only one ever discovered. But even the huaqueros were disappointed, for their predecessors had beaten them to the booty. The entire area had been dug and panned decades earlier in 1851.

The years following World War II brought education that resulted in historic enlightenment and fostered an appreciation of the pre-Colombian cultures as well as their golden artifacts. It now was unthinkable to melt down the recovered golden artwork. The artifact and collector value far exceeded the gold bullion value. As archaeological concern for the gravesites grew, the image of the huaquero, the graverobber, now became one of a dark, ruthless figure stealthily moving through the jungles carrying both a gun and a long metal probe with which to seek out the softer earth that might mark an Indian grave. Many labeled the huaqueros as criminals and implied that a perverse pleasure was taken in robbing the dead of their gold. Simultaneously, a brisk foreign market developed to purchase the "tainted" spoils. During the 1950s clever buyers made the rounds of the jungle villages offering prices substantially higher than bullion value for any and all recovered golden artwork. It was the buyers who achieved enormous profits in these years. The illiterate huaqueros remained generally ignorant of the golden artwork's true value on the foreign markets.

EL DORADO IN MUSEO DEL ORO

The same education and enlightenment that created the foreign markets for the pre-Colombian golden artwork also affected the Latin American governments. Since virtually all the goldwork was exported from the nation where it was recovered, laws were passed to restrict the activities of the huaqueros. Colombia, backed by its spectacular golden Indian heritage, led the way in 1960, with most other Latin American nations following suit. The matter of the huaqueros and their recoveries was turned over to the Colombian Bank of the Republic, which issued a controversial offer to purchase on a no-questions-asked basis all gold recoveries from the graves. The bank offered what was considered to be a fair price.

The law had a dual effect, for while it effectively stemmed the export of pre-Colombian gold, it may have actually stimulated the unscientific digging of graves. The approach, however, must certainly be considered very successful. It was based upon a basic point of human nature. Since it was no secret that the graves contained gold, laws or not, someone was going to try to acquire that gold. Success of the program was measured by the rapidly growing collection of pre-Colombian artwork in the Museo del Oro, the gold-collecting branch of the Bank of the Republic. By 1963 that collection numbered over ten thousand separate pieces of pre-Colombian golden artwork.

The Museo del Oro in downtown Bogotá, Colombia, houses the most magnificent and complete collection of pre-Colombian gold in the world—over twenty-six thousand pieces. Curators admit that over 90 percent of all that gold has been acquired from huaqueros. Much of the gold is on public display in a presentation as spectacular as the gold itself. Visitors are ushered into a specially designed display gallery in gravelike darkness. When the lights are turned on and intensified, the reaction of the visitors is invariably the same. It is one of whispered awe, as if they woke from the sleep of centuries to find themselves in the middle of El Dorado itself.

GUATAVITA LIVES

The enduring legend of El Dorado, the gilded man, received a monumental boost of credibility in 1969. About one hundred miles south of Lake Guatavita, a Colombian peasant farmer and his fourteen-year-old son crawled into a hidden cave to investigate his neighbor's report of the presence of ancient earthenware pots. The farmer did indeed find a large ceramic pot, and inside was what has since become the most celebrated piece of pre-Colombian gold in existence.

The piece was in the shape of a raft seven inches long. Intricately fashioned, it showed an Indian chief surrounded by his four priestly acolytes, exactly as if it had been made according to the description recorded by the Spanish gold seeker Freile 336 years earlier. A tiny bit of the metal was broken off and taken to the village dentist for confirmation that it was gold.

CHAPTER 23: PROFESSIONAL GRAVEROBBERS

The magnificent raft was then shown to the village priest who was an amateur archaeologist.

"For years I had prayed for confirmation of the story of Guatavita," the priest whispered, "and there it was in front of me."

24
LEGALITIES AND REALITIES

Had the pre-Colombian artwork been made in a medium other than gold, this chapter would not be necessary. From the beginning the Spanish would have respected the Indian graves simply because sacking them would not have been worth the trouble, and the huaqueros would have tended their bananas and not dug in the jungles. Entire cultures, along with their languages, customs, and craftsmanship, at least in part, may have survived, and today thousands of graves would remain as intact cultural deposits for archaeological study.

MARKET FOR GOLDEN ARTWORK

But that was not to be, for the graves contained gold.
The acquisition of gold assumed nearly automatic precedence over considerations such as respect for cultures and the sanctity of religions. Gold appealed to every culture for widely different reasons, and it was far too coveted to lie forever in forgotten jungle graves. In the United States and Europe growing ranks of wealthy collectors stood ready to purchase any and all pre-Colombian gold. They were little

concerned with cost, simultaneously voicing great concern for "cultural" considerations. The artwork was viewed as a solid investment by those who could afford it. The price of the pieces reached new levels as Latin American governmental restrictions gradually cut into the available supply. Even the lowliest illiterate huaquero held out for higher prices for his recoveries. More and more, the huaqueros were joined by foreign gold seekers, lured by rising collector prices. By 1970 several larger operations were active, hiring many peasant laborers as diggers for the going rate of a dollar a day.

THE LONG ARM OF THE GOVERNMENT

The first hopeful attempts to halt the huaqueros' booming business came as independent actions by each government, but the laws were largely hollow with no means of enforcement. By 1972 the United Nations Convention for Protection of World Cultural and Natural Heritage recognized and denounced the continuing exploitation of the graves. As with many United Nations declarations, this was ineffective.

However ineffective, the United Nations declaration inspired a second and much more meaningful agreement. This was the Convention on the Defense of the Archaeological, Historical, and Artistic Patrimony of the American Nations, conducted under the auspices of the Organization of American States and better known as the San Salvador Convention of 1976. This was a multilateral agreement with teeth. It was geared specifically toward the protection and preservation of the pre-Colombian Indian graves of Central and South America. Not only was the exportation of illegally obtained cultural objects illegal, but their importation into another treaty member nation was outlawed also, and that included nearly all Western Hemisphere nations. Terms were even provided for seizure and extradition proceedings in order to eventually return objects to the nation of origin. Latin American nations from Mexico to Peru, those possessing pre-Colombian gold-bearing graves, passed their own stiff in-

CHAPTER 24: LEGALITIES AND REALITIES   187

ternal laws prohibiting digging, possessing, buying, selling, and exporting of cultural objects from the graves. All the regulations were similar, founded upon the basic precept of government ownership of all cultural objects still in the ground. Not only did the huaqueros risk time in the miserable Latin American jails, but buyers and sellers were subject to the same penalties since the markets were illegal. According to the letter of national and international law, the five-hundred-year-old tradition of the huaquero in the Americas had ended.

LAW OF THE JUNGLE

In reality, in the jungles, it was a far different story. The establishment of the antiquities laws coincided with the dramatic rise in the price of gold in the 1970s. Just as values of specimen gold nuggets and ancient Spanish gold coins rose sharply, so did values for pre-Colombian gold, legal or not. Now even the simplest nose ring brings about one thousand dollars on a foreign market, and a three- or four-ounce piece of more elaborately worked gold might command ten or even twenty thousand dollars. Prices like that assure digging will continue.

The threat of jail is a poor deterrent to a local huaquero. With one simple piece of worked gold, he easily exceeds his annual income, which, without digging, amounts to not more than five hundred dollars. The antiquities laws do not frighten away the foreign gold seekers, either. They, better than anyone, realize the incredible prices pre-Colombian gold brings in the right markets. Both foreign and local huaqueros find further encouragement and profit now in the nongold objects in the graves. Once discarded as worthless, the pieces of jade, onyx, mother-of-pearl, and carved bone are now quite valuable in their own right.

The enforcement responsibility of the antiquities laws falls squarely upon the local police, usually those in the outlying jungle districts where a two- or three-man outpost could represent the total regional government authority. These police have always received a very modest, in fact, almost humiliating, salary. If they are not already huaqueros themselves, most will turn their backs for consideration received.

Consider for a moment the circumstances of the policemen's lives. It is their country, their jungle, and their ancestors have lived there since time eternal. In fact, it would not be impossible that a distant, early relation may occupy one of the overgrown graves the official is bound to protect. The laws he is charged with enforcing are made in a distant national capital that grants him little interest and even less assistance in his day-to-day affairs. Even a modest "gift" of one thousand dollars is enormously persuading and economically comparable to bribing a United States city policeman with thirty times that amount.

NEW TECHNOLOGY FOR OLD GRAVES

Most huaqueros still rely on the traditional long metal probe to locate the graves, probing the ground in search of softer earth that may indicate an ancient manmade excavation. Many graves were originally constructed as visible mounds, but the passage of centuries with rapid jungle growth, heavy rains, and high humus build-up have rendered most indistinguishable from their surroundings. Metal detectors are useful in aiding excavation but provide little help in actually locating the grave.

Of far greater effectiveness are the new electronic underground *utility* locators. These instruments operate on the principle of ground, or downward-looking, radar and read and record the reflections of the dielectric constants of various objects. Every material, when charged, has its own particular and identifiable constant. In practice, the ground is charged and the instrument zeroed to the ambient measurement of the soil. Conductors, such as gold, and nonconductors, such as jade or bone, will be revealed by reflections on opposite sides of the scale. In experiments, these instruments have been successful in detecting pottery at a depth of six feet and in locating hollow areas such as caves or air-filled ceramics at substantially greater depths. These new instruments are expensive, costing about ten thousand dol-

CHAPTER 24: LEGALITIES AND REALITIES

lars, but are portable and have obvious "archaeological" applications. A word of warning should be given, for customs officials are no longer naive, and a foreigner porting a trunk full of electronic equipment will not fly into Bogotá, Panama City, or San Jose without difficulty. The officials are quite aware the owner will not use them to search for pennies on his penpal's front lawn.

Let there be no question: Seeking gold from the graves or purchasing gold that has come from the graves of any Latin American nation is clearly illegal. The penalties for being apprehended, or even charged, with participating in such activities are severe. For the local huaquero in his jungle, it amounts to little more than a slap on the wrist. But for the foreign gold seeker it will be unpleasant from the beginning right through to the end when he will probably purchase his release for a minimum of ten thousand dollars.

However, the price of pre-Colombian gold on the foreign markets will only go higher, and the huge profits being made in digging and dealing the gold will insure that, for the huaqueros, it will be business as usual.

25
THE GRAVES

The *plaza central* in San Jose, Costa Rica, is lined with tourist shops, ice cream parlors, and open-fronted restaurants—a favorite place for tourists and city visitors to lounge away the afternoon and evenings after the usual tours of museums and cathedrals. Over good food and cold beer, customers will be approached by street entrepreneurs purveying everything from evening entertainment to, yes, pre-Colombian gold artifacts.

AGUILAR'S FROGS

Without introduction, a short, round-faced Costa Rican took a seat at my table and promptly informed me that it was a great pleasure to find Americans enjoying his fine city. Costa Rica was a country of many things, he reminded me, where one might acquire fine souvenirs, the finest of which was very special gold. Would I care to inspect some, perhaps?

My neutral expression was not the desired response, so my guest formally presented his business card. Beneath the gentleman's name was the word *Dealer,* and below that, in smaller print, was a comprehensive list of what was dealt including "medicines, antiquities, special orders, translating services, gems, and gold." With sidelong glances implying what would transpire to be highly confidential, he removed from the inside breast pocket of his baggy jacket a manila envelope, from which he extracted a 1½-inch-long golden frog. Nearly a thousand years old, he whispered. From the graves.

Outside of museums, this was the first purported pre-Colombian gold I had seen. The tiny casting had the authoritative weight of gold. The frog had a finish just rough enough to suggest ancient or crude manufacture. More manila envelopes appeared, and from them came more golden frogs, both smaller and larger, then, a small bracelet of crudely joined gold links.

My reluctance to purchase anything brought a long look of disappointment as he returned each piece to its envelope, then placed the entire stack in a series of inside jacket pockets that looked like a file drawer. His parting words reminded me that I could contact him, day or night, at the telephone number on the card if I should have second thoughts.

ALL THAT GLITTERS IS NOT FROM THE GRAVES

"I see you met Aguilar," said my new guest, who now filled the still-warm chair. His English was good, although laced with a French accent, and he spoke through a broad grin. "He is a character. I have known him for some time now and also know that everything he sells is fake. He works with a jeweler who makes it in his back room."

Renaud preferred to be known by the Americanized *Ronnie*. With his wiry build and hawkish nose overhanging a thin moustache, he would have made a terrific cat burglar. He had lived in Costa Rica "off and on" for several years and had learned quite a bit about the local gold. Aguilar, he explained, sold low-karat gold fakes to European and American tourists. Ronnie personally judged the work to be sloppy and flawed, but good enough to pawn off on tourists who were thrilled to acquire something "from the graves." Aguilar might even "dig" something close to a buyer's personal preference, perhaps a jaguar or serpent figurine, but such orders were available only at a premium price because the imaginary huaqueros would have to make a special effort. Aguilar sold his pieces for more than double the bullion value of an equal weight of pure gold, an excellent markup for fourteen-karat gold.

Buyers were assured of the legality of bringing the gold back to their homes but often found differently later at customs. Some declared their purchases openly and had them seized. If an inspector could determine their true origin, usually through poorly burnished modern two-piece mold marks, the innocent buyers would be informed with a huge, satisfied smile that they had been ripped off on fakes.

BIRDS OF A FEATHER

And how did Ronnie become so well versed in matters of local gold? Simple, he explained, he was a part-time digger himself. Our friendship grew quickly when Ronnie learned that I was working shipwrecks along the Caribbean reefs.

Both nursing headaches, we set out the following morning along the Pan-American Highway toward Liberia in the north. Ronnie was driving a brand-new, expensive Ford pickup truck with a temporary Texas registration taped in the rear window.

When I commented digging must be good these days, Ronnie explained that the truck would not be his much longer. To finance his digging, he would fly to Texas, purchase a new pickup for cash, and immediately drive it south across the border. At the Costa Rican border a bribe to "his" customs inspector provided the correct entry papers.

The truck could then be sold to any wealthy Costa Rican for twice what Ronnie paid for it. Those buyers would bribe someone else in turn to be able to legally register the vehicle and resell it, if desired. Everyone made out, with the exception of the government, which never collected its import duty. After his own costs and bribes, Ronnie claimed he made three thousand dollars, which he invested into his growing digging business.

"It's only right," he said. "The biggest cost I have is bribes anyway."

RENDEZVOUS WITH THE TENIENTE

About a half-hour from the Nicaraguan border, we changed from the truck to a battered jeep, then turned off the highway onto a dirt road leading west toward the Pacific. Ten miles later, the steadily deteriorating road passed through low green hills and small stands of bananas and citrus.

Finally we got to a small village of not more than twenty tin-roofed shacks. Ronnie stopped the jeep at the largest building. There was a faded Costa Rican flag nailed across the porch.

A very young policeman wearing khaki shorts and a uniform shirt and trailing a rifle emerged from the door. Seeing Ronnie, his face brightened with a wide grin of recognition. Behind him another figure appeared at the door, half undressed and obviously only a moment removed from his siesta.

"The *Teniente*," Ronnie said. "This guy is *it* for everything around here."

The police lieutenant studied me with a long silent stare while speaking to Ronnie in hushed tones. After a while he offered his hand in a reserved gesture of welcome. "The Teniente and I are partners," Ronnie said with a relieved smile. "I just told him that you and I might work together. He's quiet, but don't worry, he just wants to make sure who you are. He doesn't see new gringo faces every day."

On the way to the shack that would be home for a few days, we passed by the local jail, a sheet metal and adobe structure built into the indentation of a natural rock ledge.

"I put my time in there," Ronnie mentioned with a gesture. "Anybody can break out in a second, but nobody does. The guard, he's only a kid, will shoot you. Don't laugh, I've seen it happen. They shot the guy I was locked up with. He went out of the hole first, and I had one foot out of the

CHAPTER 25: THE GRAVES

same hole when he got it. I changed my mind quick and stayed there."

In the first light of morning, Ronnie, the Teniente, five huaqueros, myself, and a burro loaded with digging tools moved along a faint footpath past the banana trees and into the jungle. After a mile Ronnie broke the silence to explain that the overgrown mounds we passed were really heaped-up dirt left by huaqueros thirty and forty years ago.

"Lots of graves around here, some dug, some not. Some beautiful pieces came out of here years ago. There are still a lot more here, the trick is to find them."

After reaching the area he wanted to dig, Ronnie assembled the threaded sections of a long steel probe, then used it to delineate a section of soft earth found on a previous visit. The huaqueros were put to work with the picks and shovels, two working on a previously started trench, and the other three on a new hole. The Teniente made himself comfortable in the shade of a tree.

For hours the digging progressed in silence. Ronnie moved from hole to hole, checking the bottoms to see if a burial area was being reached. At midday the only others we would see all day, two machete-wielding peasants, passed by. They stopped to exchange a few words with the Teniente, then moved on.

THE WAY IT WORKS

In the early hours of the afternoon, the huaqueros put aside their tools to eat a meager lunch while Ronnie explained the arrangements of digging. Aside from a few remaining local huaqueros doing free-lance work and selling to local buyers, most digging had evolved into a system that more or less benefited everyone involved.

In Ronnie's own situation, the Teniente, who now snored loudly, provided his assurance of protection and cooperation. The huaqueros, all of whom had done their own digging at one time or another, usually with mixed success that brought a rags-or-riches payday, were being paid many times the going rate for peasant labor, whether or not anything was recovered. All had been selected and approved by the Teniente in a controlled unionlike arrangement. Ronnie himself provided the contacts with the highest paying foreign markets as well as with the persons necessary to export recoveries. At this time Ronnie was personally taking the recoveries to Managua, Nicaragua, crossing the border along a trail where "no one stamps passports," then turning them over to two respected Nicaraguan diplomats who carried them without trouble into the United States.

Ronnie had sold his first recoveries to a single wealthy individual in Miami. That person offered more than any San Jose buyer could match. Lately he had made progress in organizing a circle of affluent collectors who cared not about price at all, only supply and authenticity.

Pre-Colombian gold artwork is not illegal to possess outside the country of origin, and many of the pieces are eventually acquired by museums as gifts and donations. The collectors' "gifts" are not as philanthropic as they may appear, for each item is a full tax write-off, and prior appraisal by the right expert will assign extremely high values to each. Such values are rarely questioned by government tax agencies, for it is not easy to argue the point of lesser values for golden artwork of now-extinct cultures.

"The poor huaquero," Ronnie said with disgust. "He gets the blame for everything. He robs graves, he tears up cultures, he's no good. But what about these rich collectors outside? Man, they'd kill each other to buy this stuff. Why don't you blame them instead of these poor diggers with ten kids who can't turn down five dollars to dig holes in his back yard? Do you have any ideas how many parrots and monkeys are smuggled out of here for the States? Cat skins? But that's not so much of a terrible crime. The trouble is that we're talking about gold. It's judged differently, so it's a big thing.

"Do you blame me? The Teniente? I'm making a buck like anybody else, and the rich that buy this stuff will do the same thing. The government doesn't give a damn about these people, it's worried about the heritage.

"Heritage," Ronnie emphasized the word with a snicker.

CHAPTER 25: THE GRAVES

"These buyers on the outside with all their fancy houses and education don't give a damn about heritage, they give a damn about gold."

A GRAVE GIVES UP ITS GOLD

In the late afternoon the trench was pushed through the area of loose earth, recovering only a few scattered fragments of pottery, evidence that other huaqueros had been there first. The other hole had reached a depth of seven feet where it uncovered the edge of a stone circle. Ronnie inspected the bottom of the hole and ordered part of the wall taken down and the base enlarged to expose the grave.

At last light, Ronnie again slid down into the hole and, on his knees, began sifting through the loose earth. Now even the Teniente, who had moved little all day, peered intently over the side. Mumbling to himself as he sorted through the debris of the gravesite, Ronnie passed up the first of his recoveries, a small ceramic pot four inches high. That was followed by a few worked sonte pieces and drilled animal teeth. In the weak, yellow beam of a flashlight, Ronnie now set aside human bones while continuing his search of the loose, red earth.

The scraping stopped for a few seconds, and the only sounds in the rapidly falling jungle darkness were a few bird cries and the buzzing of insects. Finally, Ronnie stood and passed up a small, mud-smeared object. Although it was covered with the reddish clay from the grave, we could see in the flickering, failing beam of the flashlight that it was gold— a flat pendant nearly four inches high in the form of a strange stylized human figure. All I could hear in the jungle were low murmuring among the huaqueros and the sound of more scraping in the pit.

It was completely dark when two huaqueros helped Ronnie out of the grave.

"That's it, no more here. This guy was not such a big hombre," he said, looking back into the hole. "Still good for a day's work."

It was actually a very good day of digging. The pendant,

which weighed perhaps three ounces, would bring well over ten thousand dollars in Miami. Before beginning the long walk back toward the village through the pitch-black jungle, one of the huaqueros quietly threw a few shovels of dirt into the exposed grave to return the dead to their sleep.

GOOD AS GOLD

Two years passed before I saw Ronnie again in San Jose. He was happy to say the digging business was better than ever. He no longer ran Ford pickup trucks into the country and now enjoyed living in a comfortable, spacious apartment in a pleasant, residential area of the Costa Rican capital. Ronnie still had problems, political ones now, that had put an end to his Managua connections and his digging operations near the border.

"It's bad up there now," he said, referring to the rapidly approaching climax of the Sandanista revolution. "Even on this side of the border there are soldiers, everybody's soldiers, sneaking around all over."

Ronnie again expressed his regrets that I had not accepted his earlier proposal to move his recoveries by boat. In fact, it had been some time since Ronnie had made any shipments at all.

"A temporary inconvenience. One must have patience in the digging business, it is foolish to do business with the first guy that comes along. Perhaps, in another year, Managua will quiet down. Do you remember the place we dug when you were with me? Only two, maybe three hundred feet away we found some good graves. It took us a month but, well, here, let me show you."

Opening a suitcase and an attache case, Ronnie withdrew his collection, each piece wrapped in newspaper and fastened by a rubber band. When every piece had been unwrapped, the rug in his living room held a very impressive display. There were several pendants, cast figurines of jaguars and serpents, a necklace of small hollow orbs, ear ornaments, nose rings, a shallow plate, a spectacular hammered piece in a stylized sun design, and a breastplate made of linked, flat

CHAPTER 25: THE GRAVES

plates. All of it in gold. Ronnie's answer to my question of retail value in outside markets brought only a shrug. Based upon the probable value of the pendant I had seen recovered two years earlier, I silently estimated that not one penny less than a quarter-million dollars was glittering on the rug.

When was Ronnie taking it outside?

"Whenever the time is right. This stuff has been in the ground for a thousand years, so there's no sense rushing anything now. Besides, tomorrow it'll only be worth a little more. It's gold."